TOOLS

**500+ TOOLS
FEATURED**

TOOLS
The ultimate guide

JEFF WALDMAN

ILLUSTRATED BY HOLLY WALES
PHOTOGRAPHY BY STUART BALLEW AND TODD MONROE

CHRONICLE BOOKS
SAN FRANCISCO

Library of Congress Cataloging-in-Publication Data
Names: Waldman, Jeff, author. | Wales, Holly, 1983- illustrator. | Ballew,
    Stuart, photographer. | Monroe, Todd, photographer.
Title: Tools / Jeff Waldman ; illustrated by Holly Wales ; photography by
    Stuart Ballew and Todd Monroe.
Identifiers: LCCN 2021035712 | ISBN 9781797209876 (hardcover)
Subjects: LCSH: Tools--Catalogs.
Classification: LCC TJ1200 .W35 2022 | DDC 621.90075--dc23
LC record available at https://lccn.loc.gov/2021035712

Manufactured in China.

Design by Rachel Harrell.

Typesetting by David Van Ness.
Typeset in TT Phobos and Avenir.

10 9 8 7 6 5 4 3 2 1

Chronicle Books LLC
680 Second Street
San Francisco, California 94107
www.chroniclebooks.com

TABLE OF CONTENTS

INTRODUCTION

This book is an exploration and celebration of tools. It's not a complete reckoning, but it spans most commonly encountered tools and many of their lesser known variants. It's an attempt to unravel the complex tapestry of how tools relate to one another, their use, and the esoteric language that enshrouds them. Simply put, this book will teach you what a hammer is, tell you which one to buy, and explain how it came into being.

Tools are an important part of the human story. Our evolution is chronicled by the tools we've made. Stone Age. Bronze Age. The Industrial Revolution. Sure, in part it's because those tools are the remnants that survive us, waiting millennia to be unearthed along some riverbank, but it's also because our ability to shape the world around us is largely what's defined our species. We've been doing it for millions of years, and in lockstep our evolution of purpose-built tools has grown in size . . . while also becoming ever less approachable.

That's the difficulty of these storied objects. They are an intimidatingly deep well of specifications and jargon that's daunting to unpack.

Think of this book as a low-resolution map. While it might not have detailed information on *every* tool, it provides context, connecting the dots for the newly interested while filling in some gaps for the more

seasoned. By cataloging a wide variety of tools, I hope to plant the seeds for future discoverability and help you ask the right questions.

This is exactly the sort of resource I wish I had owned before embarking on my own self-taught journey of tool purchasing, tool use, and constructing a veritable homestead and cabin. It would have alleviated much wasted time and confusion.

When my partner, Molly, and I first purchased our land, it was an undeveloped bramble, with one unmaintained road dividing a steep ten acres of oak and redwood. At the time, all I owned was a cordless drill and a few other apartment-grade hand tools. I had a little experience with tools and building, but my knowledge was piecemeal and inadequate for designing and developing an off-grid property.

We started simple, with a picnic table and benches, then a firewood cubby, and soon elevated tree decks, an outhouse, and an outdoor shower. Many other small and larger projects with our friends followed, each one building upon the experience of the last. We were emboldened by each success and by the lessons learned in failure, and it showed in our evolving aspirations. Learning about some new tool or technique would open the door to discovering another, and this fostered a deep appreciation for each tool's qualities.

But our progress was also stilted. I was frequently stuck on the most basic questions concerning the right tool for the job or conventional practices, and I got bogged down under the weight of too many options and too much information. *Am I using this the wrong way? Am I losing time to the incorrect tool? How did I just cut myself?* And always I faced the aforementioned problem of finding the proverbial tool catalog daunting to unpack. *There are SO MANY tools.*

When I first considered the leap into nail guns—a tool I knew was critical to an eventual cabin build—I read up on the different types: powder-actuated, air, gas, electric. Within those categories, models were further divided by their specifications. There was a lot to sift through, but I was particularly hung up on each gun's angle. Some were 30°, others 21°, and so on. *Degrees of what? Which was better?*

It turns out that the angle is inconsequential, but questions like this stymied my decision-making for over a year because I knew that some numbers definitely mattered—mistakes can cause decks to collapse like beach chairs—but I didn't always know *which* numbers mattered.

I ground through the questions and confusion—nail guns included—taking missteps and baby steps, building confidence and competence. In time I gained an intuitive understanding of the right tool for the job and an earned familiarity with the wrong tool for the job. I amassed tools and, more importantly, the skill to use them.

When Molly and I (along with a sizable cohort of friends) took on the challenge of designing and building our cabin, it was intimidating to say the least, but when we completed construction a year later, it was no ramshackle hut. Instead, it had a polished interior with high ceilings and lots of glass and was perched on a hill with a view across the valley. We'd started down a road that was scrappy and uncertain but eventually completed a project that was featured in books and architectural magazines and garnered enough admiration that *our* expertise was now being sought. I had a real feeling of accomplishment standing on our cabin deck, marveling at all the hard-won experience and knowledge I had gained through trial and error.

But it was a prolonged path. The uncertainty and naivete that peppered all that work (and, if I'm being honest, a lifetime of dabbling with tools) made for exponentially more effort and a litany of mistakes. It took me a long time to cultivate my appreciation for tools and collect the pieces of their vast puzzle. It was a slow build with many aha realizations, enabled more by curiosity and tenacity than a proper education. I made it work, but there are easier paths. I certainly could have benefited from that wisdom up front—from a mentor or a comprehensive book—but I'm excited that now I can at least share all that accumulated knowledge with others.

This is the book that would have saved me years of slowly gleaning information, a lot of confusion, plenty of embarrassing mistakes, and probably a few trips to the first aid kit. Hopefully it'll do the same for you.

HOW TO USE THIS BOOK

You'll notice that the tools are organized alphabetically within four categories: Measure, Cut, Fasten, and Shape. This categorization is meant to help conceptualize tools by their function, but it is by no means absolute. Think of these categories as a guiding narrative rather than a strict definition of a tool's purpose.

With that in mind, there are a few ways to use this book.

First, if you're just curious, flip around. Browse at random. The entries are self-contained, and you're likely to learn a bit of history or trivia or be introduced to some tool variation you've never heard of. Many tool entries are also a vehicle for more involved exploration. For example, "Digital Multimeters" offers a primer on electricity; "Rulers" is a deep dive into how society ensures repeatable and trustworthy measurements; and "Screwdrivers" examines the various screw drive styles and how the world came to adopt them.

If you don't know much about tools, the best way to use this book is to read it front to back. Terms are defined the first time they are mentioned. Concepts and tool use tips are introduced where appropriate but not repeated for every tool they apply to. In fact, lessons intrinsic to *nearly all tools* are interspersed throughout the book.

Another way to use this book is as a reference. Tools go by many names, so the index should be your main resource when investigating a specific practice, problem, or tool.

That said, this book is not an encyclopedia. It is a broad and imperfect look at the spectrum of tools found in homes, shops, garages, and your local hardware store.

When it comes to purchasing the tools listed in this book, here's my advice: Buy the cheapest model you can get your hands on. Use it. Abuse it. Break it. Learn if it suits your needs and whether it will see frequent use. Armed with more intimate knowledge, you can then consider buying a costlier replacement to last a lifetime. Exceptions to this approach would be when a quality tool costs marginally more and likely won't be much of an experiment in ownership, like a good pair of locking pliers.

In general, you'll find that I've avoided mentioning brand names unless a truly standout tool calls for it. This is because there's often little difference between brands at comparable price points. Cordless tools are a prime example, and a shop's allegiance to a brand generally speaks more to an investment in batteries than it does quality.

Finally, the most important consideration in using this book is putting it into practice. Buy or borrow tools. Hit up your local tool library. Pick up a tool and feel the weight of it. Figure out how a tool works and how it doesn't. Make something. Fix something. Learn by doing. Otherwise, this is all theoretical, and any knowledge gained is just as easily lost.

SAFETY

You will need some basic safety equipment when working with almost all kinds of tools, and you should be mindful of the attire you're wearing.

- **Safety glasses** (generally a good idea)

- **Ear protection** if you're making noise

- **Gloves**, unless you're using a power tool that rotates and snagging a glove would be a hazard

- **Hard hat** if anyone is working overhead

- **Respirator** when producing fine dust and noxious fumes

- **Closed-toe shoes** or, better yet, **steel-toe boots**

- **Natural fiber clothing** if you're producing sparks or a flame (synthetic fibers can ignite or melt onto your skin)

You'll want to follow some best practices for the jobsite, shop, or even your apartment.

- Keep your work area clean.

- *Knolling* (a term coined by architect Frank Gehry's furniture shop janitor, Andrew Kromelow) is the artful and orderly arrangement of objects. Knoll your tools on the bench when

you have several out. Return them when not in use. This
keeps a space tidy and avoids injury from a misplaced tool.

- Police any trip hazards, especially power cords and
  small scraps of material that are likely to roll an ankle.

- Use clear communication with others, especially in noisy environments and when working
  together, such as lifting a heavy load.

- Take note of where switches and circuit breakers
  are located in case you need to quickly shut off
  the power in the event of an emergency.

- Have a fire extinguisher readily accessible,
  and keep it closer when making sparks.

- Keep a first aid kit visible, and write the address of
  the nearest hospital emergency room on it.

Above all, respect power tools. I mean *really* respect power tools.
Mind where your hands are before you power one up. The high
torque of an electric motor and the indifference of its wiring mean
that it won't even slow down while it maims you. Read all your
manuals and consult a variety of sources before embarking down
the road with a new tool. If possible, get some in-person instruction.
Professional instruction is mandatory with some of the more involved
tools covered in this book, such as lathes and welders; for these tools,
the how-to discussion here is more of an informative theory of operation than it is a guide.

MEASURE

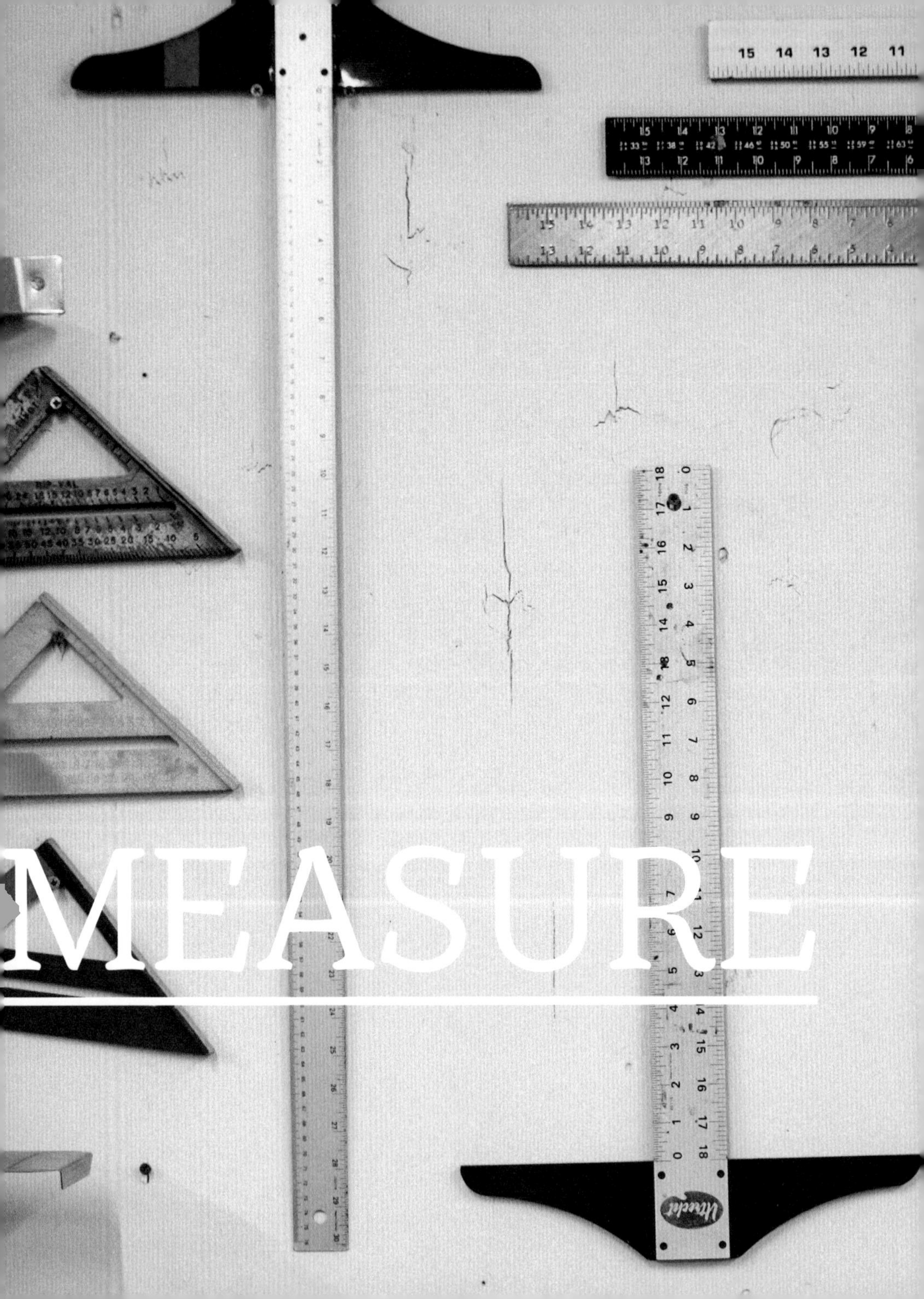

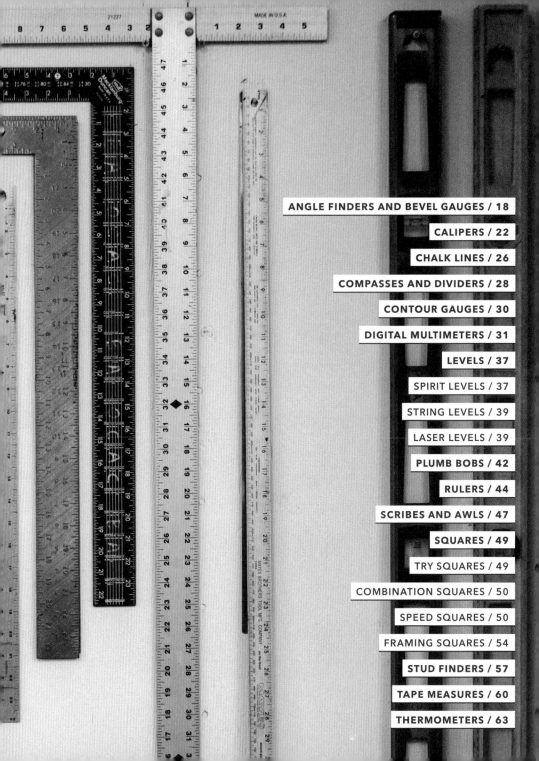

Angle Finders and Bevel Gauges

We humans have been measuring angles for about as long as we've been shaping the world around us. Take the Göbekli Tepe site in Turkey, the oldest large-scale construction project yet unearthed. The site features a twelve-thousand-year-old temple whose perfectly level, square-edged megaliths are arranged in an equilateral triangle. That's a lot of angles that needed to be calculated.

Right angles have always been the most critical, and devices were built early on to ensure things were square. Ancient Egyptians used precisely knotted rope pulled into a right triangle. (A triangle with three lengths between knots on one side, four on the other, and five on the third forms a perfect right triangle where one corner is 90° square. You can use the same trick with your **tape measure**, measuring any units of three, four, and five that you choose.) The Romans did their surveying for cities and encampments with a **groma**, a tool that resembles a child's mobile on a staff. Four plumb bobs were hung from a square cross, their swaying steadied in tins of oil. Survey lines were sighted along these plumb strings at right angles to one another, making possible the layout of perfect grids across vast landscapes.

In time other angles needed to be created and, more than that, recorded. Ancient Babylonians decided that dividing a full circle into 360 units, called *degrees*, was logical, partly because of their base 60 mathematical system and partly, it seems, because of the approximate number of days in a year. Their system was adopted by the Greeks, so in 240 BCE, when Eratosthenes measured the angle of the sun to determine the circumference of the Earth (with 98 percent accuracy!), he did so in degrees. Today we still use those 360° to measure everything from the angle of a car's driveshaft to the cradling slant of a park bench.

There are a few other systems for measuring angles. We express the slope of a roof in *pitch*, a measure of rise over run, for example. And radians are the official unit of measurement for the metric system and mathematics. However, degrees remain humanity's most common unit for measuring angles.

Angle finders are sold under a few names and shapes. The simplest ones will tell you the angle of the surface they sit on and little else. There are U-shaped **inclinometers** for leveling recreational vehicles, and pedestal-mounted dials that are sold under the same name. But the most common angle finders are magnetic indicators that sit on a surface. These are marketed as **digital magnetic angle finders**, **angle gauges**, or **magnetic protractors**. Analog dial versions can be more cheaply had, but digital magnetic angle finders are more useful as they can measure the angle between two planes, such as in setting the angle of a table saw blade, or the sharpening angle of a chisel. Other common uses are for establishing the angle of pipes and finding a specific grade for drainage. For this reason, most digital angle finders have various modes to cycle through that convert the slope between degrees, percentage of grade, millimeters per meter, and inches per foot.

You might not need to *measure* the angle; it may be enough just to be able to *duplicate* it. Recording an angle to transfer it to another medium is common in carpentry and metalwork, and the tool for this job is a **sliding T bevel**: a stout block with a nested metal blade that pivots and locks down with a finger-tightened

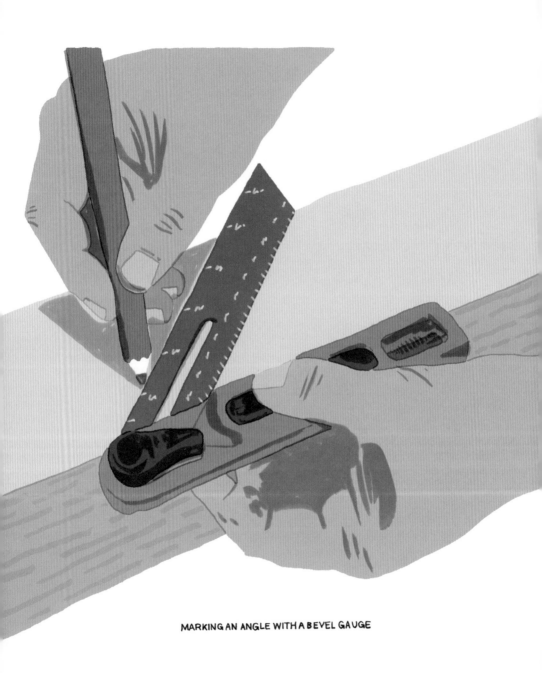

MARKING AN ANGLE WITH A BEVEL GAUGE

screw. Should you need to actually measure the angle, there's the **digital bevel gauge**, which is essentially a sliding T bevel but with a digital readout. And there's also the **digital protractor**, which has two metal rulers, joined by a pivot, and a digital display. Both of these digital tools have buttons to reverse the measurement, so that rather than reading the angle all the way around an arc, you can switch to measuring the angle between the arms.

A common use for these tools is finding and bisecting the angle of mitered corners, but that's a job better done by a **miter protractor**. Look at the baseboard or molding in your living room. Note how the two angled pieces wedge into one another at the corner. Those are mitered corners. In theory, most of these intersections are 90°, but in practice they are usually a bit off, so creating a seamless joint requires measuring the exact angle and dividing it in half before heading to the miter saw, where you would cut two pieces to fit perfectly together. But translating the angle that you measured to the angle of cut on the miter saw is surprisingly tricky. The numbers on the miter saw's own scale are not technically accurate. When making a perfectly straight cut, the scale on the saw reads 0°—but it's actually a 90° cut. This is because you're cutting straight across that piece of lumber, so the cut is 90° relative to the length of the wood.

Confusing? You bet. This is why miter protractors exist.

A miter protractor spreads its pivoting blades between the two walls to give readings that correspond exactly to the numbers used on a miter saw's scale: the number to which you'd set the saw to cut two angled pieces that would meet flush in the middle for a mitered corner. No division or subtraction required.

Calipers

Calipers use adjustable jaws to measure the exact external or internal dimensions of an object. These jaws seat squarely against whatever is being measured, making them especially useful for measuring round objects (like pipes and tubes), which are difficult to measure with a ruler.

A quick note on the difference between pipes and tubes and how to measure them. **Pipes** *are used to transport liquid and gas, are always round, and are sized according to their internal diameter.* **Tubes** *are structural. They might be round but can also be square or rectangular. They are measured by their external diameter.*

Calipers are precise instruments, generally measuring down to a thousandth of an inch or centimeter. (To put that in perspective, a human hair is about 3/1000 inch, or 7/1000 centimeter, thick.) Calipers are often reserved for machinists and mechanics who perform microscopic measurements, but their indispensability in quickly measuring the width of a screw, hole, or hose makes them useful to other trades.

Calipers are available with three different types of readout—digital, dial, and vernier—but they all operate in the same basic way.

The lower jaws are used to grasp and measure the outside of an object, and the upper points wedge within any interior space you're trying to measure. Where the jaws meet flush is the spot that's designed to take a measurement. The tail, which extends from the back, can also be plunged into an object as a depth gauge, with the frame seated against the top of whatever is being measured. When the jaws are fully closed, the calipers should read O. To ensure this, the dial on dial calipers can be rotated to O, and digital calipers can be zeroed with the push of a button.

When it comes to reading the measurements, the three types of calipers indicate a bit differently. Measurement readings on **digital calipers** are self-explanatory, and most can conveniently switch between metric and imperial units. (*Imperial* refers to the inches and feet distance measurements used in the United States and a few other countries. Other terms used for the same system are *standard* and *SAE*, which stands for the Society of Automotive Engineers.)

The readout on **dial calipers** is a little trickier. You first read the ruler along the length of the caliper, then combine that with the dial reading. Let's use a metric caliper as an example. If the jaws are open just wide enough where they pass the 10 mark, that's 10 millimeters. Then let's say the needle of the dial is halfway between .1 and .2. Adding the two together gives a total reading of 10.15 millimeters.

Vernier calipers are less common these days. They work similarly to dial calipers, but the dial is scrapped in favor of a scale invented by Pierre Vernier, a seventeenth-century French mathematician. The vernier scale was a big improvement on a rudimentary tool that had been relatively unchanged since ancient Greece, and the fine-resolution scale proved so useful that it made its way onto **sextants** (celestial navigation tools) and various other instruments that require exact adjustment. Vernier's scale is read by

CALIPER MEASUREMENT

determining which graduations seem to line up best. Depending on where exactly the jaws are open to, a single mark on a subscale will line up just so with the main scale. (Some even have an additional subscale.) By combining the value of these aligned marks, you arrive at the total measurement.

If you need to measure tighter tolerances than calipers can offer, **micrometers** are similarly built instruments that generally measure a shorter distance with greater precision. Both micrometers and calipers should never be stored with their jaws fully closed, as temperature changes and metal expansion can bend the frames and throw off future readings.

Chalk Lines

Chalk lines are used to mark a straight line between two points. Usually, a chalk line is used when the distance is longer than any available straight edge or when a rough surface isn't conducive to marking with a pencil.

What makes this so useful is that even on pitted surfaces like pavement, or across staggered fence boards, the chalk is an obvious marker that highlights a straight path. But take care if the chalk will remain visible once the job is done, as some chalk colors can't be washed off. Check the label on the package, but as a rule use washable white chalk on visible surfaces, under wallpaper, and under white paint, where brighter colors might bleed through. Red is typically nonwashable and suited for more permanent marking. Neon colors are a good choice if you're feeling festive.

The chalk line itself is a string housed within a spool with a bent metal tab tied to the end. The line is pulled out via the tab and hooked onto an edge where the line is intended to be marked. If there's no place to hook the tab, a nail can be temporarily placed, or a helper can hold it. The line is then pulled taut by the person holding the spool and lowered gently onto the surface, so as to lay the string where it should go but not yet leave a chalk mark. Then the line is grabbed as near to the middle as one can reach, plucked straight up, and released without any sideways movement. If this is done correctly, the line will snap down exactly where it was previously resting, casting a line of chalk onto the surface. This is called *snapping a line*.

The line from the powdered chalk will be fuzzy and about as wide as a line drawn with a marker. For many applications, such as laying a straight line of fasteners or sawing off overhanging deck boards, that degree of accuracy suits just fine. However, other applications, such as the tight-fitting connections of timber frame joinery, require the finer mark of an **ink line**. Ink lines lay down a permanent mark about as thick as a pen line. For increased accuracy, don't hook the line with its wide metal tab. Instead use a knife to notch each end of the workpiece (the piece of material you're working on), and slip the line into those notches, wrapping the rest of the line down and under the workpiece to hold the line in place. A line snapped this way, and with a careful pluck, will fall precisely where intended.

After you've snapped a line or two, chalk and ink lines should be reeled back into their spools to replenish the line for the next snap. The supply in each spool will last for many days of work, but eventually it will need to be refilled with chalk or ink.

Compasses and Dividers

Compasses are more commonly associated with drafting, but they've long been a staple tool for all kinds of builders. These aren't to be confused with a compass that points to magnetic north; the type of compass helpful in drafting and fabrication is used to draw circles and arcs.

Compasses have two legs—one with a sharp point and the other terminating with pencil lead. Some have legs that bend, which allows for more precise pivoting. (The point and pencil lead aim straight down, rather than at an angle.) Whatever the style of legs, they all splay out from a center pivot.

To use a compass, place the point in the center of the intended circle, and spin the pencil point around, drawing an arc. It's a simple tool, and any old compass is useful for marking on most mediums. For metal, though, it's best to use **dividers** to scribe a line.

Dividers are little more than a compass with no pencil lead. Their two sharp points are adjustable in width, and they're used in drafting to transfer and compare measurements, but in metal fabrication they're used to lightly scratch circles, arcs, or two parallel lines onto a workpiece. Coating the surface with **layout fluid** (sometimes called **Dykem** or **marking blue**) will make these scribed lines much more visible.

Another use for either of these tools is tracing the contour of an irregular object onto a workpiece that needs to be cut to fit into place. Let's say you want to shape a deck board to fit neatly against brick facade. The board is laid alongside the brick and the pointed leg of the compass is moved across the masonry while the pencil leg moves in parallel and traces the contour of the brick onto the board. The deck board can then be cut to fit exactly into place.

For jobs too large for a compass or dividers, there are **trammel points**. These scribing tips can be clamped to a piece of wood or other fixture, creating a compass of any length. Having a set of trammel points on hand can streamline a lot of projects where drawing circles of varying sizes would otherwise require searching the house for round objects to trace. If you don't require perfect accuracy, another acceptable compass substitute is tying a pen to a length of string and pivoting it around a half-sunk nail. A shoelace will do. If you're desperate enough, anything can be a compass.

Contour Gauges

Contour gauges, also known as **profile gauges**, have a row of sliding pins that can move independently but then hold their position. They're used to transfer a complex shape from one medium to another. They are not unlike the pin-art toys that were popular in the 1990s that would temporarily record the impression of a hand, face, or small object.

In the conservation fields, these tools are known as **pottery gauges** and are used to record architectural flourishes or the profile of historical vessels, but their more frequent application is in construction, where they're used to copy unique shapes that some material has to conform to.

A common example would be cutting a piece of flooring to fit around a doorway. The layers of existing baseboard, the door trim, and the shape of the entire doorjamb can make for a complicated pattern, and the easiest way to map it out is to use a contour gauge. The shape of the door frame is recorded by pressing the pins tight against the surface. Then the gauge is removed and used as a guide to trace the pattern onto the piece of flooring that's to be cut. Once cut, the piece should be a perfect match and fit snugly into place.

Digital Multimeters

Hundreds of specialized instruments can be used to test the various qualities of electrical energy and electrical components, but the most fundamental measurement devices have all been distilled into a single handheld package, the **digital multimeter**.

The multimeter was invented in 1923 by a British post office engineer, Donald Macadie, who was fed up with lugging multiple instruments to telecommunications sites. His consolidated meter evolved over the decades, and the analog scale eventually gave way to a digital display. **Analog meters** *are still used by those who find it easier to monitor oscillating levels via a needle that dances back and forth, but for the vast majority of applications, the digital multimeter's utility, ease of use, and accuracy have rendered analog versions obsolete.*

Digital multimeters can perform a variety of functions (including measuring voltage, current, resistance, continuity, and frequency), depending on the model and the cost. Here are the basics, as well as a primer in electricity.

First, you can measure *voltage* (or, more simply, *volts*), which is the electrical potential of a live circuit. A meter might be used to check for the correct number of volts in a battery or an electrical outlet. Not seeing the numbers you'd expect to see could mean that a battery is dead or a wire is compromised somewhere.

The Current War of the late 1800s was an outright battle between electrical pioneers who fought over which form of electricity would power electrical grids. It was only after decades of subterfuge and propagandist performances (which included the electrocution of a surprising number of livestock and circus animals) that AC was acknowledged to be the most efficient way to move electricity over vast distances.

The setting labeled with $\overline{\overline{V}}$ is DC voltage (short for *direct current*, the steady stream that emanates from batteries and within many electronics devices). It might also say "VDC." The setting with the $\widetilde{V}$ is for measuring AC voltage (short for *alternating current*, the undulating wave of electricity that runs through overhead power lines and inside your walls), and this might also say "VAC." Many multimeters have a range you must select from, and the labels for these ranges may be indicated with symbols for metric prefixes, such as micro (millionth), milli (thousandth), kilo (thousand), or mega (million). If you aren't sure what you're about to measure, start on the highest range to protect the meter, before moving down the scale to get a more accurate reading.

After selecting the meter's setting and range, you'll take the actual measurement by touching the metal probes at the end of the **test leads** (pronounced "leeds") to two contact points, such as the two slots in an outlet. Because electricity flows around in a circuit, it's directional. Batteries have a positive and negative side, as do outlets. If you get a negative reading, switch the leads between the two points you're touching. Black (often labeled "comm" on the meter, for "common") should go to the negative or neutral side of the circuit, and red should go to the positive or "hot" side of the electricity, but there's no harm to the meter if you accidentally reverse this the first time around.

Meters are also capable of measuring *current*. Current, also called *amperage*, is measured in amperes, or as they're more commonly called, *amps*. To understand amps, it's helpful to think of electricity in terms of flowing water. Voltage is like the water pressure, and current is like the volume of that flow. Voltage remains the same in many cases, whether it's being supplied by a battery or the electrical grid. But the number of amps that flow through the system is determined by the device that is drawing them. Appliances that create heat, such as a toaster, will draw many amps. The starter on your car, which needs an immense amount of power to turn the engine over, also draws many amps.

Multimeters that measure amps have both AC and DC amp settings, labeled with an "A," and usually two different connection points for the test leads, depending on the range of amps being measured. As with volts, start high if you're unsure, but also take note of the maximum number of amps your meter is capable of reading. Exceeding that number will break the meter or at least pop the fuse inside it.

Most meters don't measure watts directly, but it's worth covering this common term to get a complete picture of electrical measurement. *Wattage* (or watts) is a measurement of total power and is derived by multiplying the voltage (volts) and current (amps). You'll often see the energy usage of household appliances expressed in watt hours per year, which is the power the appliance uses multiplied by the number of hours it runs in a year. If you're looking to check the energy consumption of an appliance, a **digital outlet power monitor** will do what your multimeter can't.

Next up is *resistance*, another measurement function of all digital multimeters. Resistance is measured in ohms, and per the formula of Ohm's law (current = voltage/resistance), the lower the resistance, the higher the flow of electrical current. Let's revisit the water illustration. If voltage is the water pressure, resistance is the diameter of the pipe. A wider pipe, with less resistance, means more flow . . . aka more current, so more amps.

Measuring resistance differs from previously discussed measurements in that the meter is not reading the electrical current of a live system. In fact, if you're trying to measure resistance, you don't want any electricity present. Instead, when measuring resistance, the meter injects its own power into the system and looks at how that electricity flows out one lead, through whatever is being measured, and back into the other lead. Setting your meter to measure resistance requires setting it to ohms, which is indicated by Ω, the Greek letter omega. With your meter set to Ω and your test leads not touching anything, it should read "OL" for "overload." This indicates that the multimeter is reading maximum resistance. It detects no flow of electricity between the test leads because air doesn't conduct electricity all that well. In common electrical parlance, this is an *open circuit*, or just *open*. Now, touch the leads together and it will read 0Ω (or a very low number). Whatever resistance is displayed indicates the small amount that's present in the metal of the leads and probes. A reading of zero (or nearly zero) indicates a direct connection, and this is called a *short*.

Where this functionality sees a lot of use is in diagnosing electrical malfunctions. Seeing an open reading when you should see a short or some specific amount of resistance means that a wire is broken somewhere. Plus, many components have a known resistance value, from a resistor on a circuit board to the wires attached to a car's spark plugs, and checking that value can diagnose the health of those components. Likewise, various sensors on automobiles are often adjusted by monitoring their resistance values. (You might hear a mechanic talk about "ohming it out" as a diagnostic process.) But sometimes you don't need to know the numerical reading; you just need to know if there is a healthy connected path for the electricity to follow—like an unbroken wire.

For testing the connectivity of an electrical path, most meters have a *continuity* function. It's usually the setting with the ·))) on it. Just as with measuring resistance, this is for nonenergized electrical paths, meaning you read it without any outside electricity running through it. The meter injects a small amount of electricity, and if there is a good path between whatever the two leads are touching, it emits a beep. Broken connection, no beep. Simple.

Some meters also measure *frequency*, which is expressed in hertz (Hz). The number of hertz is the number of times, per second, that the waveform of AC voltage cycles up and down. Depending on where you are in the world, the AC power in your walls cycles at either 50Hz or 60Hz. If you've ever seen fluorescent lights flicker in a slow-motion video, this is because that light is literally turning on and off dozens of times every second, faster than the eye can usually notice. The buzzing hum you may hear around power lines is also the by-product of that same rapid pulsation.

Multimeters have a few other less common features. Some can measure temperature with an additional **thermocouple probe**. Others have special functionality for various circuit board components, and there are even pocket-size **oscilloscopes** for viewing electrical waveforms. Meters can get infinitely more complex, but there are also simpler electrical measurement tools that meet basic needs. Not everyone needs to actually measure the voltage or current in a system. Many simply want to know if wires are energized before they touch them. Electricity is dangerous and invisible, so a quick reassurance that it's not about to sting you is sometimes all that's required—hence we have **voltage testers**.

Voltage testers come in a few configurations. Some plug into a wall outlet. Others come with test leads, like a multimeter. But the most useful and most common are pen-size wands meant for tucking into a shirt pocket. These **noncontact voltage testers** work by detecting changes in electrical fields, which means they can notify you if a circuit is electrified without actually touching anything conductive. These tools can be waved across a lamp wire or probed near an outlet or circuit breaker and will indicate if live electricity is present. But it should be noted that they only work with AC voltage. You won't get a reading around your car battery, for example.

Levels

Levels are devices that use Earth's gravity to verify if an object is truly horizontal, plumb, or positioned at some other angle. A level can be as simple as an app on your phone, or a clear tube of water held between distant points, or it can be as complex as a laser spinning a high-powered beam across a city block.

Spirit Levels

The essential element of all **spirit levels** (often called **bubble levels**) is a bubble of air trapped in a vial of alcohol—hence the name "spirit." The vial is marked with one or more sets of lines, and when things are level, the bubble will be centered between them. These vials are mounted in housings of all shapes and sizes to meet the needs of various jobs.

Torpedo levels are tool belt–size spirit levels with tapered ends, sometimes with a magnetic side or an extra slanted bubble for finding 45° angles. **Engineer's levels** and **machinist's levels** are precisely ground hunks of metal with calibratable vials that are used to level equipment and fabricated parts.

Box levels might be marketed as **carpenter's levels** or **mason's levels**, but they're all the same basic design: a knee-high to head-high metal or plastic housing with a few vials embedded within. These will often be used as a straight edge on the jobsite. Vials in the center are used to check a horizontal surface, while vials on the ends are meant to check if a vertical face is plumb.

Post levels are less conventionally shaped and—as the name implies—are for posts. Rather than hold a level against a post you're trying to stand straight and continually move it from one side to the other, you can use these hands-free L-shaped levels, which strap on with a rubber band, to help you determine if the post is level in two axes simultaneously. **Bull's-eye levels** similarly check for level in all directions. These button-shaped vials are found atop navigation compasses, tripods, and some drills.

When it comes to using your spirit level, accuracy matters. The high price tags of levels like those made by Stabila, and the protective cases that coddle them, are entirely warranted, but there are ways to ensure an accurate reading without going into debt.

For starters, use the longest spirit level you can for the job. The longer the level, the more complete the reading, so when framing a wall, a head-high level will give a much more accurate indication of whether that wall is standing straight than a pocket level will.

Second, verify your levels. Check when you buy them, and check again when you use them. To do this, set the level on a surface and mark where it sits. Take note of where the bubble lies. Spin the level around along the mark you made and read it again. Then flip it upside down and read it once more. The bubble should read the same every single time. If it doesn't, that level needs to be retired. This works the same when checking a level vertically. You can set the level perfectly plumb against a wall and scribe a line along it. Spin and flip it around, and place it on the line you drew. It should still read plumb.

Lastly, be exact in your readings. View the bubble with one eye centered on the vial. Stare it down like it owes you money. It's not enough that the bubble is *between* the lines—it must be *evenly* between them. (If it's not, you can raise one side of the level until the bubble is centered to get a sense of how much things need to be raised or lowered.)

String Levels

String levels are spirit levels in a different package. They're pocket-size tubes with a hook on each end. Not meant to be set on a surface, these levels are hung on a tight string to find a straight line.

When you're plotting the layout of a build, stabilized stakes are usually driven into the ground—an assembly known as **batter boards**. Strings are attached to these, pulled tight, and used to mock up the lines of what's being constructed. This might be the placement of footings, fencing, paving stones, or a concrete foundation. The string level should be hung in the center of the string. Steady its bounce, but then keep your hands off while reading whether the bubble is centered.

Laser Levels

The SLAC National Accelerator Laboratory in Menlo Park, California, is one absurdly long building with a perfectly straight line of electronics: a 3.2-kilometer-long [2 mi long] particle accelerator. Leveling it in the 1990s took months, but a scientist there joked that had they waited just a few more years, it could have been done in a day with a store-bought laser level. She wasn't wrong. Laser levels have come a long way in a short amount of time, and these days they're affordable, ubiquitous, and useful for a lot more than leveling billion-dollar atom smashers.

A **laser level** is the tool to use when a span is too far for a spirit level or when you'd rather avoid the headache of setting up strings. Laser levels are great for projecting a level line across multiple spots at once, like across a row of deck posts or a kitchen wall. Many laser levels emit both horizontal and vertical lines, which is useful for aligning cabinets, trim, tile, or shelves.

Laser levels can be had for cheap, but spending some extra money in a few key areas will be well worth it. For starters, get a self-leveling model. These work by suspending the laser on a pendulum within the unit, which means that the beam that's projected on the wall is always level. Acquire one that projects a bright line (though a piece of white paper can help locate the laser of less powerful models), not just a point, and consider springing for a model that projects in a full 360° circle.

Laser levels are delicate instruments. The pendulum needs to be locked for transport, and most automatically lock when turned off, but be sure to unlock it before taking your reading. They should also be checked for accuracy. For most models that shoot a laser line in one direction, this can be done by projecting the line onto a living room wall from the center of the room and marking both ends. Then move the laser from one side of the room to the other, and aim it at each mark. The line should land on both marks. If your level shoots a laser 360°, the easier verification is to mark a wall where the laser line lands and rotate the laser 180°—just like flipping a spirit level around. The laser should still be on the same mark. Vertical lines can be verified with a hanging plumb bob.

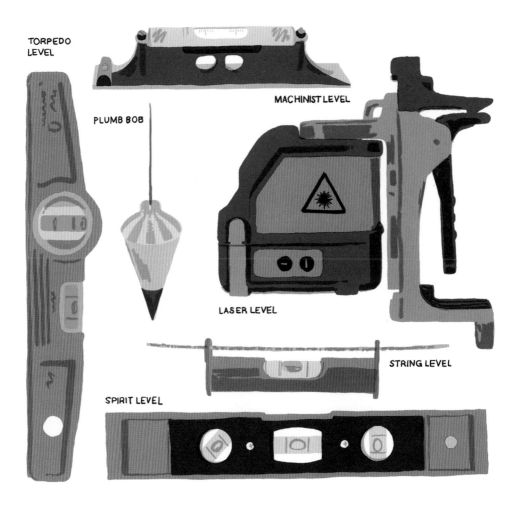

TORPEDO LEVEL

MACHINIST LEVEL

PLUMB BOB

LASER LEVEL

STRING LEVEL

SPIRIT LEVEL

Plumb Bobs

A **plumb bob** (or **plummet**) is a weight on a string; it uses gravity to find *plumb*—which is to say, a truly vertical line. It can be as simple as a rock tied with some twine, or a tape measure that's locked off, but both of those are inherently imprecise. A true plumb bob is a heavy piece of lead (the name *plumb bob* comes from the Latin word for lead, *plumbum*), brass, or steel with a hardened, tapered point and a fixture on the opposite end for tying a centered string. This construction ensures that the weight hangs true and points straight down from where it's suspended.

Plumb bobs might be used to square the framing of a wall, to find a spot in the ceiling to mount a stove pipe, or to transfer the layout of a string line straight down to the ground. Larger plumb bobs have been hung inside skyscraper elevator shafts to ensure vertical construction and from within the center of stone spires as they were raised. They have always been a foolproof method of finding a vertical line, no matter the height.

Many plumb bobs have a replaceable tip threaded on, as tips can get damaged. The other end also unscrews so that the string can be threaded through a centered hole and then tied. They must be steadied once hung, which can be surprisingly difficult on a long length of string; heavier plumb bobs help with this. They should also be free of any obstruction. For instance, if you're using a plumb bob to see if a wall is plumb, don't hang it so close to the wall that it touches anything that might interfere. Instead, hang the string a short distance away from the wall, and then measure between the wall and plumb bob. And try not to suspend it from a string that has knots along it, as those can throw off a reading.

Laser plumb bobs are an option if you're not big on twine. They consist of a small plumb bob in a pendulum fixture that rests on the floor, projecting a laser beam straight up. These are useful, accurate, and not terribly expensive. But because lasers are basically magic, they should be periodically checked against a traditional plumb bob.

USING A PLUMB BOB

Rulers

A **ruler** seems pretty simple. It's a flat piece of metal, plastic, or wood, about the length of your forearm, with markings for inches or centimeters laid out along the face. In most cases, it contains 12 inches, or 30 centimeters, though shorter and longer ones are available. (In fact, there are extra-long **folding carpenter's rulers,** and some people prefer their reassuring rigidity to tape measures.) Sometimes rulers are used for measuring a certain length or drawing a line to scale, but just as often their edge is used just to make straight lines. They're not complicated tools.

However, wrapped up in those hash marks is the history of humans taking measure, the standardization of those measurements, and the regimented verification of measuring tools against an official standard.

Consider the trust we put in our measurements. Rulers are useless if one ruler has a longer inch or centimeter than another. For large-scale construction projects (and really for society to exist at all), consistency and trust are required. You can't have a crew cutting half of the blocks for the Great Pyramid of Giza with one set of measurements and another crew using another. This dilemma has always been further compounded when you consider that some projects (like said pyramid) take decades to complete. That's a lot of time for workmen to scribe their own rulers (called **story sticks**) and for the marks on those sticks to get smudged. Transcribed measurements can wander over time, much like a game of telephone.

Ancient Egyptians recognized that risk with their own version of the ruler: a unit of measure known as a *cubit*, which was about half a yard or meter in length. Their solution was to implement the first system of standardized measurement and a calibration schedule that ensured traceability and accuracy. Every full moon, wooden field cubits were to be checked and logged against *the* cubit—meaning the royal stone cubit. The punishment for failing to do so was death.

Fortunately, those draconian measures have fallen by the wayside, but that first system of standardized and traceable calibration lives on. What happened in Egypt was the birth of the field of metrology (the study of measurement itself) and the foundation for the centralized entities that govern it, such as the National Institute of Standards and Technology (NIST) and the International Organization for Standardization (ISO).

Today, most of us take for granted that a gallon of gasoline really is a gallon, that a kilogram of bananas truly is a kilogram, or that a car's speedometer isn't just guessing. Society hums along with the quiet agreement that we're all operating within a shared reality, and behind the scenes the system that makes this possible has remained largely unchanged since the pyramids were constructed.

Gasoline pumps are verified annually using a calibrated container, which itself has been checked against an even more accurate vessel by some other independent authority. Every step and piece of equipment is logged, on schedule. The scales at the supermarket weighing those bananas are certified with weights four times more accurate than the scale's manufacturer claims it to be, and those weights are themselves periodically checked against an even more accurate scale. On it goes. We're surrounded by these measurements, from the calories on a food label to the signal emanating from a cell phone tower. These are accurate and the technology is functional because of a system of standards and calibration that creates a traceable chain that can be followed back all the way to *the* standard of measurement.

When it comes to the measurement of distance and the modern ruler—the descendent of that first cubit—standards have looked remarkably similar to the system of old. From the late 1700s until 1960, the official meter was defined by a securely stored length of platinum metal, just as it was in Egypt with the stone cubit. (Similarly, the foot was long defined by a bronze yardstick in a Washington, DC, vault, but since the scientific community was doing such a bang-up job with the metric system, in 1866 the foot was redefined as a fraction of that platinum meter.) Accurate copies of the platinum meter were sent around the globe, and of course sent back for periodic checks. Those standards were used to calibrate exponentially more standards, and those were used to ensure that the veracity of the ruler you purchased never had to be questioned.

This system of periodic checks against more accurate standards has remained consistent, but *how* standards are defined has evolved. In the mid-1900s, metrologists started to worry about how much of a bind we'd be in should a fire melt that platinum bar, or if tragedy should befall any of our other physical standards. So they started redefining units, from physical objects to constants in nature. In 1960, the length of that metal bar was measured one final time, and the meter was officially redefined as the distance that light travels in a vacuum in $\frac{1}{299,792,458}$th of a second.

In many ways, that feels more complicated than a hunk of metal kept under lock and key. More ethereal. *Less trustworthy*. But it's accurate, and it's repeatable. And it means that even if there is a total societal collapse, or a fire in that vault where we store the meter, the kilo, and all the rest, we can rebuild, and what once was an inch will always be an inch.

Scribes and Awls

A **scribe** (or **scriber**) is a tool used for etching a line onto a work-piece. It's common to use a stray nail or a knife, and of course there are pencils. But rather than dull your knife or scrawl a wide line that might smudge, a sharp scribing tool is often the most reliable choice. This tool is especially important in metalwork, where scribing over **layout fluid** is the clearest method of marking the hard surface.

Metalworking scribes are pen-shaped tools with a sharp tip of some ultra-hard metal, generally carbide. Those **carbide scribes** can etch aluminum, steel, glass, and even stone and ceramics. While they tend to be marketed toward metal fabrication, it should go without saying that they're also quite capable of scratching marks into wood.

Scribing in woodworking is often done in place of a pencil to create a finer line to follow or to mark a divot to start a drill bit or fastener. This might be accomplished with a tool marketed as either a scribe or a **scratch awl**. Using a scratch awl to lay lines for cutting has the added benefit of scoring the grain, which will produce a cleaner saw cut. However, if this is the main goal, a **marking knife** is better suited for the task, especially if the line is cutting across the grain. But the fine points and long tapers of scratch awls are useful for all kinds of other tasks: etching soft sheet metals, picking at carpet seams, reaming out small holes, or aligning two holes with one another. The tool can be heated to poke cauterized holes in synthetic fabric or reshape the crimped ends of a cable housing (like those found on bicycle brakes) after it's been cut to length. Scratch awls are also used in leather-crafting shops to trace patterns onto leather, though in leatherwork, as with canvas, a **stitching awl** might be more appropriate since these can also be used for feeding stitching through holes. Stitching awls are identified by the hole in the spike, and in fact the overlooked, stubby little blade with the hole in it on your **pocket multitool** is a stitching awl.

A **bradawl** is shaped somewhat like a scratch awl but is flattened at its tip to produce a chisel edge. In addition to marking lines, they were intended to be used to notch the fibers of wood to allow screws and nails to be inserted more cleanly, but they're not often used these days.

While it's fallen out of popular use, the bradawl does have a lasting legacy in the invention of Braille, the tactile writing system used by the visually impaired. In 1812, at the age of three, Louis Braille was playing with his father's bradawl and accidentally stabbed through a piece of leather and into his own eye. That injury, and the subsequent infection to his other eye, cost him his vision before he was five. The disability motivated Braille to invent his alphabet of raised dots, and by fifteen he'd worked out a prototype for what we use today. In a bit of narrative poetry that Louis was hopefully able to laugh at, he happened to form that prototype using the very tool that took his vision: an awl.

Squares

Squares are named for their purpose of checking or creating a square angle—that is, an angle that is 90°. When the ancient Egyptians invented them, a square did little more than make a right angle. Since then, they've evolved into specialized tools that feature all sorts of neat functions and are often the go-to tool for marking a straight line, measuring short distances, or verifying that two surfaces are flush with one another.

Try Squares

So named because it's used to test an angle's squareness, or to "try" it, the **try square** is one of the more basic and less dynamic squares. In metalworking these might also be called **machinist's squares** or **engineer's squares**. A few companies, such as Woodpeckers, sell models with extra bevels that help seat the square firmly on an edge, along with the promise of longer lasting precision—and many pros swear by their value—but you'd need a few years' experience before noticing the difference between these and the most commonly available try squares.

Your basic try square is composed of a metal blade and a metal or wood stock. The stock is thicker than the blade so that it can be snugged up against one edge of the object, while the blade lays perpendicular atop the surface.

Combination Squares

The **combination square** was invented by Laroy S. Starrett in 1877; nearly 150 years later, the L. S. Starrett Company is still making some of the most reliable measurement tools available. Laroy's dynamic design was a big improvement on other squares of the era, and the basic shape hasn't changed since then. These squares are composed of a ruler and a handle, which is also known as the anvil, head, or frame. The ruler's position can be adjusted, or it can be removed entirely. As its name suggests, a combination square offers options.

The shape of the handle allows the user to easily check a 90° angle or a 45° angle. Some combination square kits include a protractor head that allows the user to create any angle. These kits also often incorporate a triangular head used for finding the center of a circle.

Because the ruler is adjustable, it can be used as a depth gauge from the face of the handle. This adjustment also allows for the neat trick of sliding the handle along the length of a workpiece with a pencil held on the end of the ruler, marking a parallel line at a set width from the edge. Some models have a level built in, and many include an often-overlooked scribing tool that's threaded into the handle. Rather than reach for a pencil, unscrew the sheathed point and scratch a line onto whatever material you're working on.

Speed Squares

Speed Square is actually a brand name, but you may hear it used for other brands of square. Swanson Tool Company invented the Speed Square in 1925 and marketed the tool to North American

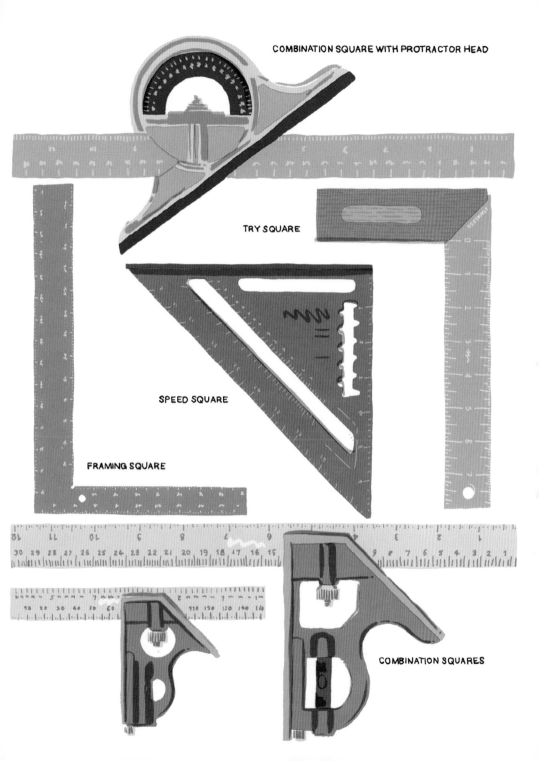

COMBINATION SQUARE WITH PROTRACTOR HEAD

TRY SQUARE

SPEED SQUARE

FRAMING SQUARE

COMBINATION SQUARES

carpenters as a "Try Square, Miter Square, Protractor, Line Scriber, & Saw Guide." They weren't wrong with the name: It's quick to deploy, easy to use, and has sped up plenty of jobs.

Speed squares are triangular, which confuses a lot of newcomers who are asked to "grab that square." That triangle shape, with a lip on one side, allows for an easy check of 90° and 45° angles. One common use is in framing, where you use a tape measure to place marks along a length of wood for stud locations and then use a speed square to quickly draw square lines across the face of that wood at each mark. In Canada and the United States, these tools are 7 inches across (though larger models do exist), so you can mark two pieces of dimensional lumber, such as 2x4 studs, at once. Metric speed squares are usually 25 centimeters across— wide enough to span side by side lumber that's 90x45mm. But they are less common because stick framing is less prevalent in countries that use the metric system.

Stick framing—the colloquial term for building structures from wood studs—is not an international standard. Far from it. Brick, concrete, and even metal studs lead much of the world's construction, but in North America, where lumber has historically been plentiful, stick framing dominates (this type of construction does appear in other locations around the globe, notably in Australia and the boreal region of Europe). That said, in any country, the size of a speed square allows it to be tucked into a back pocket, or into a tool belt, and they have enough useful features that anyone who works with wood would do well to test-drive one.

A speed square has notches meant to hold a pencil tip while scribing a line down a length of wood. At the corner of the right triangle and the end of one side of the lip is the square's pivot point.

Pivoting it there, you can mark angled lines using the protractor scale along the long edge. On the same edge is a scale for common roof pitches and rafter cuts, as well as marks for cutting the bird's-mouth notch that holds many rafters in place. All these roofing features are why some call these tools **rafter squares**. Most of these squares are 3/16 inch [5 mm] thick—a useful gauge for a quick gap check when installing a window in a building's framing.

Possibly their handiest alternative application is acting as a circular saw guide to get a quick square cut on a workpiece. Here's how to do it: Set the shoe of the circular saw on the wood you're about to cut, with the blade positioned where the cut will be. Hold the square firmly with one hand against the wood, with the straight

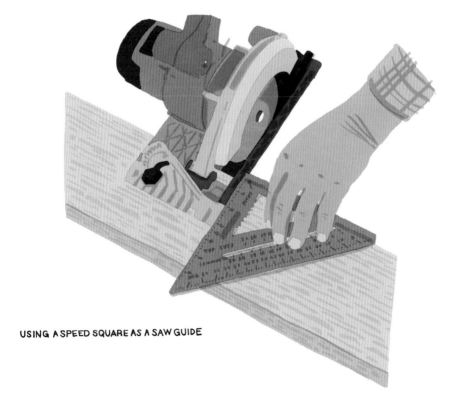

USING A SPEED SQUARE AS A SAW GUIDE

edge nestled along the shoe of the saw. Then, with those fingers clear, run the saw forward, holding it against the speed square as a guide. Do it slowly. Do it safely. Maybe watch a video. Most importantly, watch your fingers. Be careful and you'll get a quick square cut and save a trip to the chop saw.

Framing Squares

Sometimes called a **steel square**, a **framing square** is deceivingly simple. It looks to be little more than an L-shaped ruler, but there's a ton of utility hidden within . . . provided you and your framing square are in North America.

As previously discussed, stick framing is not a global standard, but it reigns supreme in North American construction, along with the imperial system of measurement. Even in Canada and much of Mexico, where the metric system is more frequently used, feet and inches are often the de facto system when it comes to framing a home. And while there are metric framing squares, they are an anomaly, so much of this deep dive on the tool will be focused on the market that it was designed for.

The short arm is called the *tongue*. It's 1½ inches wide—the width of most dimensional framing lumber—and useful for marking the outline of where studs will land on other wooden members. (Metric framing squares oddly have a 40mm tongue, even though metric stick framing lumber is often 45mm thick.) Framing squares are 16 inches long, which is the standard spacing for studs in those countries that measure stud spacing in inches. The long arm (called the *blade*) is 2 inches wide, which combines with the tongue to get 3½ inches—another standard dimensional lumber width. The blade

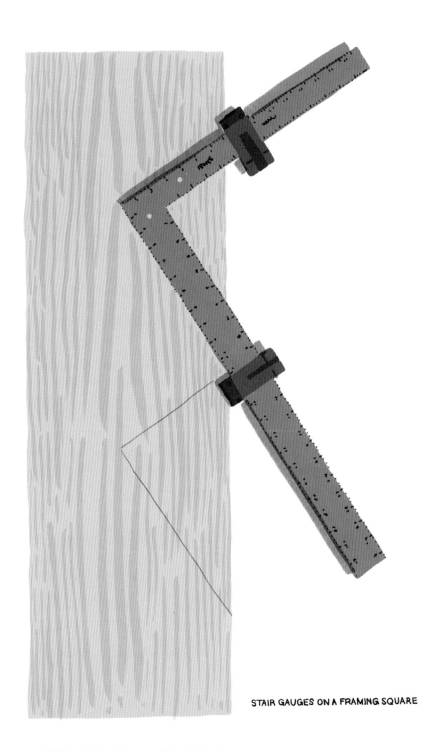

STAIR GAUGES ON A FRAMING SQUARE

is 24 inches long, which is yet another stud spacing span for a type of advanced framing and pretty close to some metric stud spacing of 600mm.

There are a couple of different scales on each side of these squares. The hash marks of conventional measurements are easily recognized. Then there are more perplexing graduations in tenths, twelfths, or even hundredths of an inch, which are used for working from scaled building plans. Measurements start on both the inside and outside of the square, so it's important to check which scale you're reading from. Confusing them could result in an error of a couple of inches.

A popular use for framing squares is for laying out stair stringers–the diagonal lumber that cradles stair steps. The step's rise (height) and run (length) are counted off on both the square's tongue and blade, and the square is used to trace a right-angled notch onto the stringer. This is repeated for each step along the length of the stringer, and those notches are cut out. **Stair gauges** are an add-on product that will save time during this process. These clamp along the framing square's tongue and blade, allowing the square to be held against the stair stringer to quickly make repeated layouts without remeasuring.

Stud Finders

There are a few types of **stud finders**, but they all aim to perform the same function: to identify lumber or other objects hidden behind walls. They also afford a seldom-missed opportunity to crack a joke about male virility.

Generally speaking, if you're using a stud finder, it's to locate the hidden studs behind a wall so that you can more securely fasten something, such as a TV mount or earthquake straps for your prized saltwater aquarium. It's crucial to know the tool so that you can trust that your fish are safely anchored and not attached to some flimsy drywall. To understand a stud finder's indications, you first need to know what's behind most walls.

Studs are vertical pieces of lumber (or metal), and in the United States and a few other countries, they are 1½ to 2 inches wide and typically spaced 16 inches apart. In the few countries that frame with studs and use metric, they're usually 45mm wide and spaced 600mm apart. Sometimes they are doubled up, especially around windows and doors. There will always be studs at the ends of a wall and usually along one side of where outlets and light switches are mounted as well. Aside from studs, there's also the possibility that plumbing, conduit, or wire is hiding back there.

Magnetic stud finders, the cheapest version of this tool, are little more than a couple of high-powered magnets. They require no batteries, won't look deep into the wall for metal objects like conduit, and will only latch onto screws and nails just under the paint. If you can find those fasteners and take note of a few, mapping out the center of the studs is pretty simple.

Edge-finding stud finders are probably the most common type of stud finder, but also the most misunderstood, as these ambiguous tools don't do much more than beep and flash a light. Here's how to use one: Place the stud finder flat on the wall in a spot that's likely not to have a stud. The best way to do this is to select a spot likely to have one, then move over a little. The sensor will be looking for materials denser than where it was first turned on, so it has to power on over an empty section of wall to calibrate. Press and hold the button. Then slowly slide the finder sideways in one direction until it beeps to indicate an edge. Mark it with a pencil. That's one edge of the stud, but you should find both edges to accurately identify the center. Release the button, move it over several inches beyond your mark, press the button, and slide it back in the direction of the mark you made. When it beeps and flashes, mark there as well. That's your other edge. These marks should be a couple of fingers' width apart and represent a vertical stud. A more advanced version of this type of tool is a **multisensor stud finder** with an array of LEDs. These will illuminate over the full width of the stud, rather than only identifying its edges.

Models with a deep penetration mode can locate dense objects like metal pipes. This feature can be helpful, but only if you know what you're looking for; trying to discern what's going on behind a wall is tricky, and there's always the risk of drilling into a pipe you thought was lumber. One way to test before you commit to driving a large anchor is to first drill a small pilot hole just deep enough to clear the drywall or other material. If you hit resistance that produces wood shavings in the drill bit, that's a stud. If you encounter a void, or a hard object like metal, you need to patch that hole and probe elsewhere.

The other way to get a complete picture of what's hidden behind a wall is to use more expensive **stud finder phone attachments**. These can differentiate between pipes, wires, and wood and show them on your phone's screen with surprisingly accurate x-ray vision.

EDGE FINDING STUD FINDER

Tape Measures

Tape measures are deceptively simple. It's all right there in the name: a tape with which to measure. But like many simple tools, there's a surprising amount of nuance and dexterity wound up in that spool.

Most tape measures are a reel of U-shaped metal tape, wound on a spring inside a case. Keychain-size models will span a table; palm-size tape measures will measure a house. For longer distances, there are larger hand-cranked **fiberglass tape measures**, stick-pushed **measuring wheels**, and **laser distance measurement devices**. (Laser measuring tools can be useful for short distances as well, especially when working solo.)

On a regular tape measure, the tape is pulled out, a measurement is taken, and then the spring returns the tape back to the case. Many models have a lock button to keep the tape from retracting. The tape is flexible, but not entirely so, as its curved profile is meant to give it stability. If you need a truly flexible tape, look for a **tailor's tape measure.** That's the simple side of tape measures.

Here's where they get more interesting: Tape measures have a hooked metal tab on the end of the tape. You may notice that it slides in and out very slightly and wonder why your tape measure was so poorly made. However, that wiggle is by design. The tab moves the same distance as its own thickness, canceling out its width, meaning that measurements made by pushing against the outside of it, or by hooking on the inside of it, always read the same. That hook also has a nail grip slot that can be used for suspending the tape measure, or for pivoting from a center point to mark a circle. That same hook is often serrated so that it can be used to etch a mark on wood or other material.

Making inside measurements with a tape measure, such as along the inside of a windowsill, can be tricky. A lot of folks will press the hook into one corner and then bend the rounded tape into the other corner, estimating about where it measures. There are two other ways to do this. First, most tape measures will say how long the case is, so the whole thing can be placed within the sill, from the back of the case to the metal tip of the tape. The second and probably most accurate method is to measure 10 inches or centimeters (or any short, easy-to-remember distance) from one side

When it comes to scribing a line with your tape measure, there's another way to draw a long one, rather than using the serrated hook. Let's say you want to cut 10 inches off a sheet of plywood. Grab the tape measure at the 10-inch mark with your thumb and forefinger, and press that forefinger against the edge of the plywood. Then hold the end of the tape with the other hand, with a pencil in that hand pressed against the hook. Slide both hands evenly down the length of the plywood, using your finger on the edge as a guide and drawing with the pencil in the other. You'll end up with a straight line down the length of the plywood, 10 inches from the edge. This is effectively the same trick you might do with a combination square or with the purpose-built notches in a speed square.

and make a mark, and then measure from the other side to the mark and add the two measurements together.

Most tape measures have some extra markings on them for convenience. Metric tapes have meters and decimeters highlighted. On imperial tape measures, traditional stud and joist spacing is usually marked by red digits. Some tape measures use diamonds to annotate appropriate floor truss spacing, but this is an infrequently used relic from the turn of the twentieth century.

On the topic of antiquated practices, if you happen to be on a desert island devoid of technology, a tape measure can be used as a **slide rule** for subtraction. Fold the tape over and bring the tip to the number you intend to subtract from, then look down the tape at the number you are subtracting. The remainder will be laying alongside it on the opposite side of the tape.

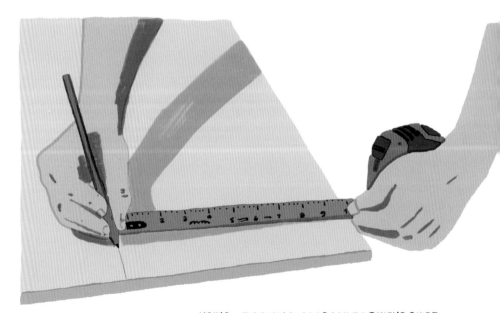

USING A TAPE MEASURE AS A LINE DRAWING GUIDE

Thermometers

There are a lot of thermometers out there, but when it comes to toolbox thermometers, the type most commonly used is the **infrared thermometer**. They might seem like specialized trade tools, but they are quite useful, and the surprisingly affordable price tag doesn't hurt.

You can use an infrared thermometer to check the temperature on a refrigerator, diagnose engine trouble in a car, and check the heat of a BBQ. Most folks will get their money's worth in the first week of gleefully measuring the temperature of every object, person, and pet in the house. It's a fun tool.

These squat gun-shaped devices are sometimes called **laser thermometers** because of the laser dot they project to assist with aiming. They work by measuring the invisible thermal radiation coming off a surface, and they can do so from a distance with no direct contact. However, the further away the object, the less focused the measurement. Up close, the sensor might be reading an area the size of a coin, but across the room that area is as wide as a dinner plate.

One shortcoming is that reflective surfaces, which have lower emissivity (the measure of an object's ability to emit infrared energy), can give inaccurate readings. Look for a matte surface, or place a piece of masking tape over the reflective area where you need to take a measurement.

SKILSAW®

CUT

Bolt Cutters and Cable Cutters

Bolt cutters are handheld tools used to cut metal that is too thick for wire cutters. They're often a speedier solution than a hacksaw. Their stout chisel blades apply enough pressure to cut through soft metals and to fracture harder ones. This extreme force is made possible by compound levers coupled with the leverage of long arms, which multiply your average squeeze into literal tons of force. When the object being cut can't stand up to that force any longer, it generally won't cleave gently in two; instead, it will give way violently. Shards of tough-to-cut objects like bolts and padlocks can be fired off at high speed in any direction. Wear eye protection when using bolt cutters.

A bolt cutter's capacity is mostly a factor of its length. Well-made extra-large models will stand belly-high, and though difficult to wield, they are capable of cutting rebar; large, hardened padlocks; bolts as thick as a finger; and fingers as thick as a finger. Smaller arm-length models will cut cheaper padlocks, medium-size screws, and fencing. The **mini bolt cutters** made by Knipex are a standout product. These German-made cutters are exceedingly efficient, and the pocket-size models—hardly larger than a pair of pliers—will cut materials that might ordinarily require a midsize traditional bolt cutter. They'll slice through chain-link fence, nails, and any metal of similar thickness.

Bolt cutters are capable of cutting some metal cable (or *wire rope*, as it's often called), but it depends on the thickness, and it helps if the wire is contained within some sheathing. Ultimately, though, they're ill-suited for the job; the jaws of bolt cutters are straight, and because of this, the strands of a cable will squeeze out from between them. If you get the cable cut—and that is a big "if"—it will likely be frayed from the process. **Cable cutters** are what you need for this job. Cable cutters have curved jaws that contain the cable as force is applied. All those twisted wires, likely to go any number of directions once squeezed, are held tight as the jaws clamp closed and eventually cleave the cable in two.

Smaller one-handed cable cutters look a lot like pliers. Larger two-handed models look like a cross between bolt cutters and the **loppers** that gardeners use to snip off branches too thick for **pruning shears**. For extra-large cables, there are hand-pump-powered **hydraulic cable cutters**, and for jobs requiring more cuts than your hands can handle, there are **electric cable cutters**. However, an **angle grinder** with a **cutoff wheel** can replace either of these specialty tools and make quick work of the job.

Deburring Tools and Countersinks

When metals and plastics are cut or drilled, they're left with sharp edges, and often a rough fringe, known as a *burr*, is left hanging along the cut. Most of the time, you'll want to remove that burr and maybe put a slight bevel on the edge so it's not so sharp. (This bevel is sometimes called a *chamfer*, and the process of dulling the edges of a material is known as *breaking the edge*.) This might be done for aesthetics and safe handling, or it may be required to ensure a smooth fit between interlocking pieces. Which tool you use comes down to personal preference, the material, and the degree of bevel desired.

Swiveling deburrers are pen-size wands that have a replaceable rotating head. Their hooked tips come in a handful of cutting angles and shapes and will put a light chamfer onto metals and plastics by shaving away a bit of the edge at an angle. The swivel motion and curve of the blade allow the cutting edge to self-align against the corner, so there's no need to be concerned about exact placement. Just hold the hook against some edge and pull the tool along it, removing wispy curls. That rotating head also makes this an ideal tool for deburring drilled holes and pipes. Even clean-cutting tools like pipe and tube cutters will leave a burr on the inside of the pipe that can foul a connection. These tools will clean it right out.

For plastic pipes, there are a couple of specialty tools. The first are pocket-size **pipe deburrers,** which are tubes that house a tapered blade on one side and the bladed interior of a cone on the other. These are meant to be used by hand and will both remove burrs and can put a fairly heavy taper on the interior and exterior of PVC pipe, ABS pipe, PEX pipe, etc. For larger plastic pipes, there are **drill-mounted pipe chamfer tools**. These are cones with blades on the inside and outside and a spindle protruding from both ends. The tool is mounted in the drill to deburr the interior of the pipe, then remounted backward to deburr the exterior.

Reamers are tapered hand tools with straight flutes meant for deburring. They're usually used on pipes but might be called upon for any hole that needs cleaning up. Confusingly, if you're in a machine shop, there are actually two types of reamers. Both are long rods that resemble drill bits with straight flutes, but only one of them is tapered. The machine shop **straight reamers** are used to machine a hole to a precise diameter after drilling the hole with a less exact drill bit, but the **tapered reamers** found alongside them might be also used in the more familiar way—cleaning off the burr in a drilled hole and putting a slight chamfer on the other-wise sharp corner.

Countersinks are reamer-like tools, but they're shorter and have a broad flare. They too can be used to deburr with a light twist, and handheld models (which look like malformed screwdrivers) are ideal for this. They'll clean up drilled holes in wood, plastic, and metal. But the real purpose of a countersink is to cut a deeper and wider bevel than a reamer or a swiveling deburring tool is capable of, so that screw heads can nest within and cleanly sit flush on the surface that was drilled. Those screws are now said to be *counter-sunk*. Countersink bits for your drill are sold alongside other drill bits and screwdriver bits.

At the end of the day, all of these tools perform the same task of putting a softer angle on an edge. It's a process fundamental to the cutting and modifying of all materials. Countersinks are just fat reamers. Reamers are merely a different shape of pipe deburrer. And the interior bladed cones of those pipe deburrers are little different than **tenon cutters**—a woodworking rotary tool used for tapering the ends of dowels. It's easy to get bogged down in all the terminology, along with *what* they cut and *how* they cut, but the basics of what these tools accomplish is universal—an important lesson to carry into much of tool exploration.

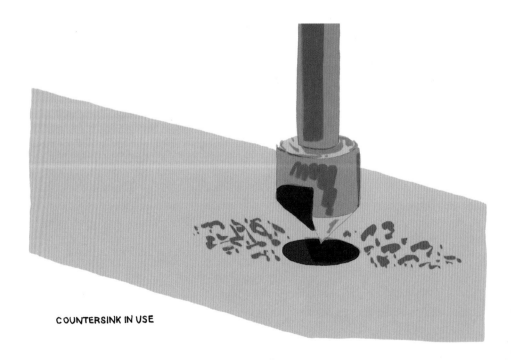

COUNTERSINK IN USE

Drills

Drills got their start as tools used to bore holes into wood, leather, and stone, though these days they're also used to drive fasteners and spin a variety of custom attachments.

The earliest models were little more than stone awls, worked back and forth with a twist of the wrist. Eventually, someone in ancient Egypt (who was clearly on the fast track to promotion) got the bright idea to fashion those awls to a stick and rotate the stick with a bow. That design proved so efficient that **bow drills** were still being used in China until the turn of the twentieth century. (Using bow drills also happens to be one of the easier ways to produce fire with friction, by spinning the wood drill in a wooden block amid some delicate tinder.)

The next iteration of the drill was the **hand-powered auger**, which more or less looked like an oversize corkscrew. The shape of these early metal bits, with their sweeping flutes that spiral down the length of the shaft like a staircase, is still seen today. In fact, the original auger, with its T-shaped wooden handle, is still used by a few artisans who relish the deliberate process of carefully augering a hole by hand, though most of the world has moved on to drills that offer speed, precision, and mechanical advantage.

Braces

Drills gained mechanical advantage in terms of torque (the measure of the force imparted on an object rotated about an axis) when the

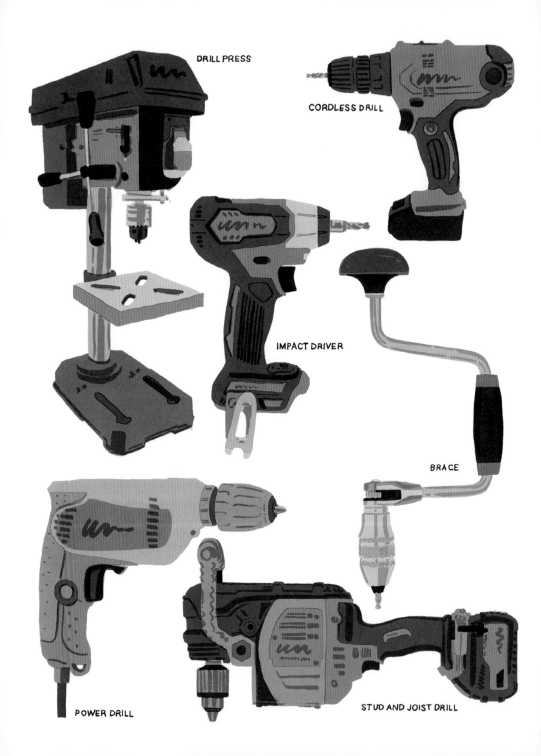

DRILL PRESS

CORDLESS DRILL

IMPACT DRIVER

BRACE

POWER DRILL

STUD AND JOIST DRILL

brace was invented during the Middle Ages in what's now France. A C-shaped pivoting device, the brace allowed the user to steady the center of the drill with one hand and to crank the handle around with the other.

Many folks would find a **three-jaw brace** to be a useful tool. These have an adjustable chuck and can be used with all modern drill bits, hexagonal driver bits, and socket adapters. When it comes to driving large lag bolts into wood, you might quickly find that your cordless drill isn't up to the task; a brace is never out of battery power, and a large one with a 10-inch handle sweep is never low on torque. Look for braces made in France. They got their start there, and that's where the few remaining high-quality models are still being manufactured.

Electric Drills

The **electric drill** is older than you'd likely imagine. This is mostly because folks tend to underestimate just how long electric motors have been in use. (The first electric scooter was cruising down boardwalks in 1915!) So how old is the electric drill? The very first electric drill was patented in 1889, and Black & Decker brought the familiar pistol-shaped drill with a trigger switch to consumers in 1917. Even their battery-powered cordless model came on the scene in 1961, again earlier than most would expect.

Since those early days, portable electric drills have become staples of the shop, garage, and home. An apartment with hardly more than a tape measure and screwdriver is likely to have a power drill stowed in a cabinet. They're common and familiar but also barely understood by many who have acquired one for a single project and never spent the time really getting to know it.

All drills have a grip and a variable-speed trigger that's meant to be squeezed with an index finger. Near that trigger is the switch that determines whether the drill is spinning forward or in reverse. The spinning front of the drill that grasps the bits, drivers, and other attachments is called the *chuck*. You can adjust the diameter of the three jaws within the chuck by turning the *chuck collar*, though it'll save time to hold the collar in place and gently squeeze the trigger. If you put the drill in forward, the jaws will cinch down; put it in reverse, and the opening will enlarge. These days, most drills have a collar that is tightened by hand, but some still use a geared **chuck key** to tighten the collar. In either case, once the jaws are wide enough to slip in your drill bit, tighten the collar down firmly.

Corded drills and battery-powered cordless drills operate in basically the same way. Though one distinction is the price you pay for the power delivered, often even the cheapest corded drills will provide more torque than many battery-powered models. Cordless drills that deliver immense torque can be had, but you'll pay a premium. Price aside, features and general performance are consistent across both styles of drill. One final consideration when selecting a cordless drill (or any cordless power tool) is whether to opt for a pricier "brushless" model. These are more efficient with battery use and won't require replacing the electrical contact brushes in the motor, which is a maintenance that might otherwise be performed several times over the life of a tool.

One quirk to note is that the jaws do not promise a perfect fit on a round or hexagonal bit. Look down into where the bit is sitting, and confirm it's centered. If the bit has flat sides, be sure the three jaws are landing on them. Smaller drill bits can also be caught between the sides of the jaws and become uncentered. If the bit wobbles when it spins, it's not secured correctly.

Many models of drill have an extra collection of settings tucked right behind the chuck collar. There's often a second ring, usually referred to as the *clutch*. The actual clutch feature will be a collection of numbers, sometimes demarcated with a screw icon, because driving screws is where the clutch is most useful. These numbers represent an amount of force that the drill will apply before the motor will spin freely and drive no further. The lower the number, the lower the amount of force. Let's say you are driving a series of deck screws. Without the clutch, you'd speedily drive a screw in, then you'd ease off the trigger as the screw head approached the surface of the decking, being careful not to over-drive the screw into the wood. This is time-consuming. Rather than doing that process for every screw, with the clutch you can find the setting (through a little trial and error) that seems to sink the screw to just the right height and breeze through the work.

For drilling, you generally want full speed and power. This means rotating the clutch collar to the drill bit icon. If your drill bit won't penetrate and you hear clicking, check the clutch to see what it's set to. Chances are it accidentally got knocked from drill mode to some low clutch number.

Some clutch collars have an additional setting to enable the **hammer drill** function. It's usually noted with a symbol of a mason's hammer, and for good reason. This setting combines the drilling with a pounding action that helps drive masonry bits.

The last bit of nuance to these drills is the gear selector, or as it might otherwise be referred to, the *torque adjustment*. Many drills have a switch on top that allows you to select between first, second, or sometimes even third gear. These selectors allow you to trade speed for power. Just as with a car, first gear is slower but has more torque. Second gear allows the drill bit to spin faster, but it will drive with less power. So, first gear might be more useful for driving larger fasteners, and second might be more useful for drilling, but it just depends on the specific application.

In situations that call for more torque than a normal drill can deliver, there are **joist and stud drills**. These right-angled beasts are shaped to fit in the tight gaps between wood framing. They pack enough slow churning power to spin large-diameter **auger bits** and **hole saws** for cutting plumbing and electrical pathways. Because of this capability, they're frequently used by timber framers, log cabin builders, and arborists, who use them to hog out mortises and to auger holes into trees. Corded and cordless models are available, and neither are to be taken lightly. Use the included secondary handle, keep a good stance, and don't hook your thumbs around the handholds. If the bit snags during rotation, the drill will violently spin away from you, taking any hooked hands and vulnerable ankles along with it.

Drill Presses

Drills are prone to wobbling and inaccuracy, which is acceptable for many situations and can be mitigated with pocket-size **drilling**

jigs to help guide the drill work. But in machining or furniture building or any repetitive drilling task requiring control and precision (especially during the low-speed and lubricant-intensive process of metalwork), a **drill press** will often produce a better outcome.

A drill press (also known as a **pedestal drill** or **pillar drill**) is a drill motor and chuck mounted to a column, where the chuck can be raised and lowered with the pull of a lever. This setup might be small enough to sit on a bench, or it might be a freestanding shop fixture that's taller than the human using it. An adjustable table sits below the drill head and can be set at a height that places the workpiece just below the drill bit. Before turning on the machine, first make certain that the drill bit won't penetrate through the workpiece into something it shouldn't. Then, by pulling the control lever with one hand, press the drill bit down into the workpiece. When pressure is released from the lever, the drill bit automatically raises back up to the top position.

Aside from the ease of one-handed operation, another benefit is that the control lever allows for a measured amount of force. Many materials require a delicate touch, frequent lubrication, and low speed to prevent heat generation. Drill presses allow for slow, steady pressure and the control of being able to retract and replace the bit into the same hole as needed.

Drill Bits

Drill bits are cutting tools used to bore holes into a material. These holes are almost always round, though there are rarely seen specialty bits for drilling square holes and other shapes.

Drill bits come in a variety of styles, but they all share some characteristics. For starters, they're all made of a hard metal and have some sort of cutting tip. Behind that tip is the body of the drill bit, which gives it some length and allows for the extraction of the material that's being cut away. Most often this is in the form of spiral-shaped flutes that swirl around the body, but in many drill bit designs (such as **spade bits** and hole saws), there's merely a void, so that the shavings have somewhere to go. If it weren't for these designs, drill bits would immediately be fouled by their own shavings, and it wouldn't be possible to drill holes of any depth.

Behind the body of the drill bit is the shank. This is the part that's designed to fit into the machine that's doing the drilling. The drill bit might be inserted into the chuck of a drill or brace or into a lathe or an automated computer-controlled machine. Drill bit shanks might even slot into a T-shaped wooden handle that's powered only by a twist of your wrist.

Bits are designed to cut specific materials. Read the packages, understand the fundamentals, and know what options exist, and you'll have a much more pleasant drilling experience.

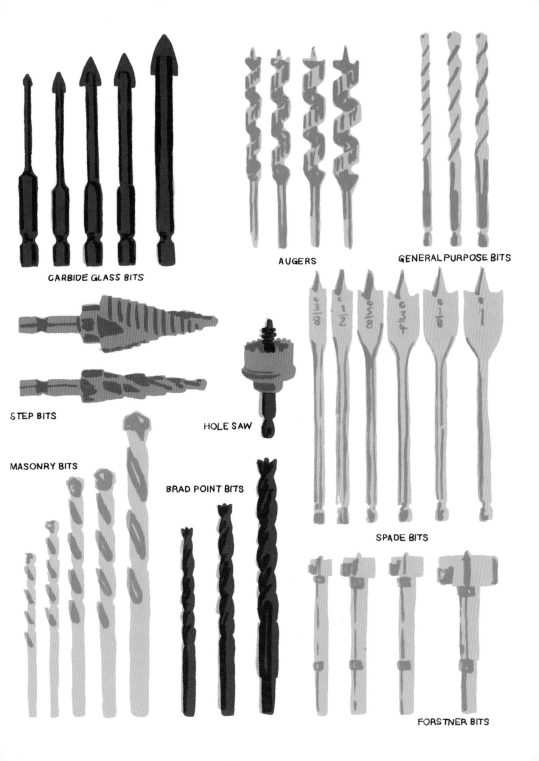

CARBIDE GLASS BITS

AUGERS

GENERAL PURPOSE BITS

STEP BITS

HOLE SAW

MASONRY BITS

BRAD POINT BITS

3/8 1/2 5/8 3/4 7/8 1

SPADE BITS

FORSTNER BITS

Multipurpose Drill Bits

Multipurpose drill bits can be purchased individually, but more commonly they're sold in sets reminiscent of a box of crayons (sometimes called *indexes*) that include the most common sizes. Indexes can vary in size and might include doubles of frequently used (and frequently dulled) sizes.

You'll see bits touting different coatings, such as black oxide or titanium. These are hardened layers meant to preserve the cutting capability of the bits. The coating of the bit can dictate the material it's best suited for. For example, titanium carbonitride is well suited for stainless steel, cast iron, and aluminum, though it will also happily chew through wood and plastics. Some consideration should be given to these specifications, but in general, if a multipurpose drill bit claims to be sufficient for wood, plastics, and metal, you need only be concerned about drilling through hard metals.

Wood-Boring Bits

Spade bits, also known as **paddle bits**, look like a canoe oar with a centering spike protruding from the tip of the paddle. They're often substituted for multipurpose bits for one of two reasons: Either you need a drill bit that will reach farther than a regular drill bit, or you need to drill a hole that's wider than a regular drill bit and don't wish to spend the money on a more robust, large-diameter bit. Spade bits are an affordable and merely adequate solution for many jobs. Perhaps the only task at which they excel is drilling in long spaces that require the wiggle room of their flimsy shank, such as creating a hole through several tight wall studs to run an electrical wire. This makes them common in electrical work, along

On Drilling

When it comes to drilling metal, it's important that the drill bit not wander. Some drill bits address this issue by having a starting point. These are sold as **pilot point drill bits** or **brad point drill bits**. (The latter are traditionally wood bits, but manufacturers now market them toward metals as well.) These points can help, but it can still prove difficult for that first twist not to spiral off the mark on hard metals. To manage this, give the bit a starting point by using a **spring-loaded center punch**, which will strike a divot exactly where the center of the hole should be. Then start drilling.

When drilling metals, go slow and use **cutting fluid**. Cutting fluids (also known as **cutting oils**) are specialty oils meant to lubricate the cutting process and keep things cool, but any motor oil will do in a pinch. Without cutting fluid, the workpiece and drill bit will get hot, and the drill bit will quickly dull. A good indicator that you're going too fast or not using enough oil is if the shavings start to smoke. For larger diameter holes, it might help to first drill a smaller pilot hole.

Drilling into wood or plastics with a multipurpose bit is more straightforward. Both can benefit from the use of an awl or center punch to create a firm starting point. And as with metal, if the hole is to remain exposed, touching it up with a countersink or deburring tool will leave a cleaner edge. The side where the drill bit exits tend to splinter, but you can prevent this by clamping a piece of sacrificial wood against the surface.

with similarly shaped **fishing bits** (otherwise known as **installer bits**), which have a hole in the drill bit for looping small-gauge wire through and pulling back through the hole, like a large sewing needle or stitching awl.

Forstner bits will outperform spade bits and are available in a wide range of diameters. They're capable of leaving a nearly pristine flat-bottom hole, which can be important in trades like furniture making. However, it should be noted that within this category of bit there are finish bits that have a delicate centering spike, and self-feed bits, which have a corkscrew pilot spike that will aggressively pull the whole bit forward.

As covered, general-purpose drill bits have helical flutes for moving shavings out of the hole, but this design only works for short distances. So what's the drill bit of choice for boring a long hole through a lot of wood? The auger! An **auger bit** is the one you reach for when you need to drill a hole through a large piece of timber . . . or through an entire tree. The tightly wound channels are, in essence, a **screw conveyor**, which is a mechanical device used to transport loose materials, like moving grain to silos. Large **gasoline-powered augers** use the same design to bore post holes in dirt and ice-fishing holes in lakes.

Hole Saws

Hole saws are so named because, unlike traditional drill bits with a couple of cutting edges, hole saws have a number of saw teeth arranged around the rim of a drum. These bits are used for punching all the way through wood, plastic, and metal, but it should be noted that they are only capable of drilling through a material that's no thicker than the drum is deep.

Hole saws require assembly. The shank and a centering drill bit are a single piece that fits into the drill, and various sizes of hole saw drums are threaded onto the shank. These can get quite large and are also available in a variety of specialized tooth patterns.

Using the hole saw is not unlike using a regular drill bit. In part this is because protruding from the center of the drum is a general-purpose drill bit. This exists to center the saw and give it something to pivot around, as well as to draw the saw into the material. You drive the drill bit into the center of the hole and brace as the teeth of the drum bite into the surface. You continue applying force until the drum clears the backside of the workpiece. Once the saw is through, pull it back through the hole.

If you're looking for that disc of wood you just cut out, it's probably inside the hole saw. That's common. The drum has slots in it for that reason. Insert a screwdriver, or a nail, and work the piece of wood out from within the drum. Often this doesn't work as well as the designers hoped, and the easiest thing to do is to unthread the hole saw from the shank and pop the wood out that way. Either way, be sure to clear these pieces, or they'll stack up within, and your drum will be filled with wood scrap and capable of cutting through nothing.

Step Bits

Step bits (or **unibits**) don't look like your usual drill bits. Their cone shape flares so broadly as to be nearly unrecognizable as a drill bit. This design allows a single step bit to drill a variety of diameters. They can be used on wood or plastic, but where they really excel is in drilling sheet metal.

Imagine you're under a cabinet, in tight quarters, tasked with drilling a hole in a metal box to pass a pipe through, and you don't know the exact diameter of the pipe. This is the step bit's time to shine. Its stout shape is ideal for narrow spaces. And you don't have to drill one hole, test-fit the pipe, then inevitably swap out the bit for a larger one. With the step bit, you instead drill a hole, pushing the cone shape as far as required to achieve the desired diameter, then attempt to slide in your pipe. Doesn't quite fit? Hit the hole again with that same bit, pushing a little farther to enlarge the hole to the next size up.

Glass Bits

It might not seem possible to drill through glass. After all, an errant pebble or mild shock can shatter it. However, drilling into glass or tile is entirely doable with the use of specialty bits.

The first style of glass bit is the **carbide spear-pointed bit**. These bits don't feature any of the familiar spiraling that would draw them into the hole and cause a fissure. And being carbide, they're harder than your average drill bit. (In fact, if you ever break off a screw or drill bit within a workpiece, carbide is your solution for retrieving it. Use a **carbide drill bit** to drill a hole in a broken-off screw or drill bit, and then feed a reverse-threaded **extractor bit** into the hole.)

Before using a carbide glass bit, first create a well around the drill site with putty or tape. Then fill that well with an ample amount of cutting lubricant. Most any oil will do, and even water will suffice. Drill at a medium speed, applying little pressure, allowing the drill bit to slowly grind away at the glass or tile.

The second style of glass drill bit is a **diamond hole saw**. These diamond-coated abrasive drums are the bit of choice if you're trying to bore a larger diameter hole than your carbide bits are capable of. These are the smaller cousins of **diamond core drills,** which are used to excavate large-diameter tubes in stone and masonry. To use a diamond hole saw, take the same approach as with the carbide glass bit. Use plenty of lubricant. Don't rush it. And soon you'll end up with a pristine hole through a brittle material that seemed impossible to penetrate.

Masonry Bits

Drilling into concrete and brick requires a unique set of tools. At the very least, it necessitates a **masonry bit,** but depending on the job, a whole new drill might be required.

Masonry bits work by turning a hard tip in combination with a hammering function. These bits can't draw themselves into their hole the way wood bits do, so they must work through a combination of pulverizing shock and rotary abrasion. A cordless drill with a hammer drill function (as indicated by a small hammer icon) can certainly get the occasional job done. But for those who frequently need to drill into masonry materials, there are dedicated **hammer drills** and **rotary hammers** that are designed purely for decimating stone.

Masonry bits with a single cutting head and a smooth shank are common and cheaply had. These will fit into your cordless drill, but they cut slower than more robust counterparts that feature quad cutting heads and a variety of shank shapes meant for locking into various types of rotary hammer. When it comes to those latter bits, it's critical to match the shank of the bit with the hammer drill it's attaching to. Several incompatible systems coexist on the shelves of retailers: Spline, SDS-plus, SDS Max. If you buy the wrong drill bit, it won't fit your drill without an adapter.

The actual process of drilling through masonry and concrete is fairly straightforward: Use medium speed, medium pressure, and the correct bit. (You may assume, correctly, that this is generally good advice for drilling of any type and material.) The most notable difference between this material and others that have been covered is the fine dust that's produced. You should of course always wear eye protection when drilling through any material, and a respirator isn't a bad idea, especially with some hardwoods. But with concrete and brick, a respirator is mandatory. Use vacuum dust collection if possible, and don't breathe in any of that mess you're making.

Glass Cutters

Just to get it out of the way, **glass cutters** are poorly named. They don't cut glass. They score a line that creates a path for a crack to follow as the glass is broken.

They come in a few shapes and sizes. **Pistol-grip glass cutters** have an arched plastic handle. The more basic models are pen-size metal wands with a ball at one end and notches which are used to grip and snap off glass, like the **grozing irons** of the Middle Ages. Then there are **circular glass cutters**, which are similar to a compass (or trammel points), with a suction cup in the center and the cutter on the other end. All these are just different housings for the V-shaped wheel that does all the scoring.

The reason these wheels are able to score a line is that they're harder than the glass they're being ground into. The uncompromising hardness of the pointed wheel applies immense pressure that forces a superficial crack along the length it's rolled. There aren't too many materials harder than glass, which is why most decent glass cutters use tungsten carbide for their wheels—the same metal used in emergency window breakers. Cheaper models get away with a hardened steel wheel, though it will dull after a handful of uses. In a pinch, the ultra-hard porcelain shards of a smashed spark plug will even work, which is why car thieves use them to shatter auto glass.

To cut glass with a glass cutter, you'll need gloves. Ensure the glass isn't tempered, like you'd find in a sliding glass door or car windows—it will shatter instead of breaking along the line. Clean the glass, and lay it on a flat surface that has a little cushion. Felt or a thin towel on a tabletop will do. If you're using a steel wheel cutter, you'll want to lay a bead of cutting oil along the length of where you intend to score the glass. Oil is less necessary for carbide wheels, but it can still help, especially on very thick glass.

If you're cutting a straight line, use a straight edge. Run the glass cutter along it one time, from one edge of the glass to the other, applying firm pressure. This should make an uncomfortable screeching sound and etch a continuous rough line across the pane. From here there are two ways to finish the cut. The first is to use **running pliers**. Pinch these specialty pliers at the start of the score, and apply just enough offset pressure to start a crack, which will then propagate along the full length of the score almost instantly. The other option is to flip the pane over (scored side down) and line up the score with the edge of the table. Press the glass flat on the table with one hand and snap it down with the other, breaking it along the length of the score all at once.

Glass cracks fast. How fast? About 3,000 miles per hour [4,800 kmh], or about five times the speed of a passenger jet.

Cutting glass with a circle cutter is a little trickier, but the same basic rules apply. Place the suction cup in the center with the cutter length adjusted to the desired radius. Score the glass in a single firm pass, then flip pane over. You now need to work a crack around the score. The easiest way is to use running pliers, assuming they can reach past the edge of the pane. If they can't reach, many pistol grip cutters have a rounded edge that's perfect for the job. Roll the pistol grip (or, if need be, the tool or knuckle of your choice) over the back of the score, causing a short run of a crack. Move along it and keep walking the crack around the length of the score. Once you've made it around once or twice, flip the glass over again, and repeat the process until you're certain there's a complete rift all the way around.

The circle won't pop out; you'll need to break the glass around it. To do this, score the external glass into four sections, then use the walking-the-crack method (or running pliers) to complete the splits. With the exterior glass cut into four sections and removed, you'll have your circle.

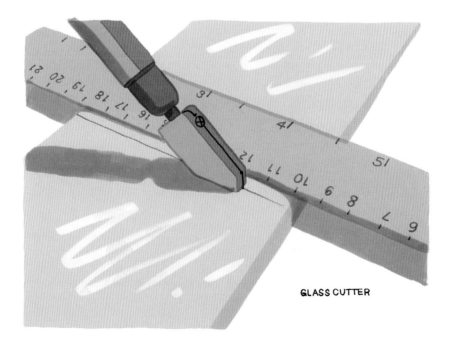

GLASS CUTTER

Hatchets

Axes were once integral to the entire process of timber construction. They were used to fell trees, shape logs, frame homes, and perform the finer points of woodworking. They've largely been replaced with more capable tools, but a **hatchet** still might have a place in your shop.

Hatchets are small axes meant to be hefted by one hand. There are no agreed-upon criteria beyond that. Some Japanese hatchets have a straight blade. Most other blades are curved to varying degrees. Utilitarian models are often sold as **camp hatchets** and come in a few lengths. **Carpenter's hatchets** are slightly larger models used in traditional woodworking and log-building, and these feature a deeper handhold for choking up nearer the head. But despite these differences, they're all more or less the same—a short length of handle with a head.

What are hatchets used for these days? All the usual roles they've filled over the years: splitting firewood at camp, skinning game, or being gleefully hurled into wooden targets. And they're the preferred tool for splitting wooden shingles to size. But where they remain indispensable is in working with freshly cut wood, known as *green wood*.

The way that lumber used to be produced—prior to **chainsaws** and mills—was by hewing logs with a **hewing ax**. A tree was felled and then scored along its length with a series of ax strikes. Then the wood between the scores was chipped away, split along the grain, creating a flat edge. This process was repeated across all four sides until a square beam was produced: a hewn log. It was the easiest way to quickly remove large volumes of wood to roughly shape a fresh log into a piece of timber.

Done on a small scale, the hatchet is an ideal tool for the job. Those who are turning green wood on a lathe, or shaping it for carving, first want to rough the wood out with a hatchet, more or less hewing it. This could be done with a saw, but rip-sawing along the length of the wood grain is slow going. It's easy enough to drive the hatchet into the wood and force a quick split along its length. This also becomes important when shaping strong wooden dowels, such as pegs for a timber frame. The grain needs to run the full length of the peg, and the only way to ensure this is to reduce the wood by splitting it along the grain—a process called *riving*. (A saw cut would inevitably drift off the path of the grain, causing the grain to "run out" along the length, producing a weaker peg.)

A quick introduction to ax lingo. The handle is the *haft*, or *helve*, and the head is said to be "hung" rather than "mounted." The blade of the head is called the *bit*, and the flat back of the head is called the *poll*. And there's an important thing to know about the poll: It's generally not hardened. If it is, the manufacturer will say so. This matters because a nonhardened poll will deform if hit. If you're using a hatchet (or any ax) to split a stubborn log and it's wedged in there, you might be tempted to strike the back of the head—the poll—with a hammer. You also might be tempted to use the poll *as a hammer*. In both cases, unless your ax has a hardened poll, this abuse will eventually ruin it.

Hatchets are also the perfect tool for the initial work that comprises most of the wood-carving process. Bowls, spoons, spatulas, and all the rest first start out as green logs and must be riven and roughly shaped until they're ready for a carving knife. Hatchets can make quick work of this process, so long as they don't send you on a detour to the first aid kit.

Of course, all this is predicated on your hatchet being sharp. Start with a reputable hatchet of quality steel. If it was manufactured in a Scandinavian country, that's a good start; this region really values a good axe. And if purchased new from a well-regarded manufacturer, it will likely come with a shaving-sharp edge, but it will eventually need tending to.

SHARPENING

Sharpening a blade is the act of refining the edge with abrasives to the finest apex possible. This doesn't just make a blade more effective; it makes it safer. A sharp blade requires less force, which in turn allows for greater control. Imagine trying to cut a piece of rope with a dull knife. The rope is folded over the blade and pulled back with one hand as you strain, sawing the blade back and forth against the cord. When the rope finally gives way, your hand and knife will spring forward, possibly into the thigh of the rope salesman standing next to you. Conversely, a razor-sharp blade can effortlessly slice that same piece of rope, with little pressure and no sudden surprises. Sharp blades are safe blades, and here's how to keep them that way.

First, know that the principles of sharpening are universal. There are varying techniques and tools for each shape of blade, but if you understand the fundamentals, you can get the job done on anything. However, it's also an art and one that's done microscopically and by feel. This primer will enable you to sharpen any bladed tool, but you'd do well to learn the finer points of sharpening by taking one of the classes commonly offered at knife shops. We'll cover the broad strokes with knives before moving on to sharpening other edged tools covered in this book.

KNIVES

Your knife likely starts out sharp, and you would do well to maintain that edge before letting it deteriorate to something that requires a true resharpening. This maintenance is called *honing*, and it involves removing the microscopic *burr* (a feathery ribbon of rough metal) that forms along the edge as the knife sees use. That rod that came with your kitchen knife set is a **honing rod** (also called a **steel**). A few swipes with the rod on each side of a kitchen knife after you clean it will keep the edge sharp. Honing rods are made of a material that's harder than the knives they're honing, so the steel rod included with most kitchen knives is satisfactory only for those it came with. For the harder steel of Japanese kitchen knives and for many pocketknives, a **ceramic honing rod** is often required.

A more refined way to maintain an edge is with a **leather strop**. These strips of leather are loaded with abrasive compounds: a coarse compound on one side and a finer grit on the other. To "strop" a knife, place it on the course side of the strop, with the narrow bevel flat on the leather and the spine of the knife tilted up at an angle. Then, trailing the edge, swipe the knife across the strop while moving the point of contact across the blade from heel to tip. Repeat on both sides and again with the fine grit. This will bring a lightly used blade back to a sharp edge.

To truly sharpen a knife that has been worn dull, you'll need a set of coarse-, medium-, and fine-grit **whetstones**, which is a broad term for sharpening stones that includes **diamond stones**, **Japanese water stones**, and **oil stones**. Japanese water stones generally produce the best edge, so use those if possible. (The process is effectively the same whether you're using any whetstone or even sheets of sandpaper.) Soak the stones in water, and be sure to keep the surface wet while sharpening. Put the coarsest stone on a towel or purpose-made **sharpening stone holder** to steady it. Line up the knife diagonally across the stone and at an angle so that the beveled edge lies flat on the stone. This will put the body of the knife blade about 15° to 20° off the stone. Then draw the knife across the entire length of the stone, while also moving across the length of the blade from heel to tip. If you get the movement down, you'll be distributing wear across the entire face of the stone while also sharpening the full length of the knife's edge.

The most important part of this process is maintaining that angle. A good trick is to use a permanent marker to mark the beveled edge and then check to see that you're rubbing it away evenly. **Guided knife-sharpening systems** assist in holding the knife and stones at a consistent angle, but since blades can have varying shapes, practice is still required.

Hold that angle and keep working one side until there is a consistent finish across the bevel, then flip the knife over and repeat. You now have a uniform edge, albeit a rough one. You should have started to form a burr, which you can probably feel with your fingernail if you push outward from behind the edge. It's an invisible little snag that's present on one side of the bevel and not the other. If there isn't one, you need to keep going. The burr is evidence that you've sharpened all the way to the edge's apex on both sides.

You'll now start the process of removing that burr and refining the edge. To remove the burr, draw the knife lightly across the stone in a straight line from heel to tip. Running the stone along the length of the burr will remove it, like grasping the branch of a bush and running your hand down it, stripping off the leaves. Do this a couple of times on each side. Then move on to the medium-grit stone and repeat all the previous steps. Once you've sanded away the rough finish of the first stone, replaced it with a medium finish, and removed the burr, you can repeat all these steps with the finest stone.

Strop your knife once you're done to remove the smallest bits of burr that remain and to polish the edge to shaving sharp. Be sure to maintain the edge with frequent stropping, as the diligent maintenance of a few swipes on the strop can save you from having to do a full resharpening for many months.

Finally, try as you might to use the full face of your stones while sharpening, they will eventually develop wear patterns, and these depressions will inhibit sharpening at a consistent angle. A **flattening plate** (also called a **lapping plate**) will level them, but coarse sandpaper and a flat surface will work just as well.

CHISELS AND PLANES

Chisels and planes have a straight blade, which means that there are jigs for sharpening them that greatly simplify the process. Don't even bother attempting these freehand; instead, use a **honing guide**. A honing guide is a clamp-and-wheel device that holds the blade at a consistent sharpening angle. Better-quality versions, like the one produced by Lie-Nielsen Toolworks, are well worth the money, but even a bargain-bin model will produce a better edge than no guide at all. Beyond this, there are also **power sharpening systems** for straight blades that rotate an abrasive disc while holding the blade in place.

To sharpen a chisel with a honing guide, you'll use the same Japanese water stones that you'd use on a knife. Mount the chisel in the honing guide. The amount of chisel protruding out the front of the guide determines the angle that is being sharpened. You'll want to match whatever is already on the chisel, which is about 25° to 30°. If you want to set your own angle, 25° is good for a primary bevel—that is, the wide slope that takes up most of the chisel's edge. And a steeper 30° is good for the secondary bevel—the narrow strip of edge that's further refined. Many honing guides have measurements labeled on the side that give a protrusion distance to establish these angles. But you can also use a magnetic angle gauge to set the angle. However, if you intend to make a habit of sharpening chisels and plane blades, it's best to build a **protrusion stop** from a few pieces of plywood. Once you know the distance that the chisel protrudes to create a given angle, glue and screw a plywood block onto a plywood board, so that next time you can simply hold the honing guide and chisel against it to set the appropriate distance and angle. Be sure to label which stop is which angle.

Using the guide to keep the bevel flat, work the chisel up and down the length of the stone as you would a knife. Apply more pressure when dragging the edge than when pushing forward. Do this until there's a consistent finish across the whole bevel. Then remove the chisel from the guide and place it diagonally across the stone, flat on its backside, and lightly slide it up and down the stone a few times to remove the burr.

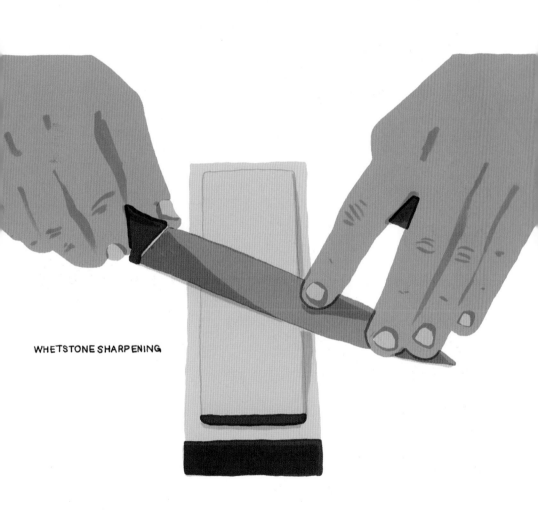

WHETSTONE SHARPENING

Repeat the process with each stone. After that is complete, you can set a new angle and add a secondary bevel if you so desire.

A light freehand stropping after this process will further refine the edge, and for maintenance you may find that a pass on the secondary bevel across a fine stone (along with stropping) is all that's required between full resharpenings.

HATCHETS

Hatchets (and their larger look-alikes, axes) can see a lot of heavy use and run-ins with rocks, so they might require significant touching up. A completely dull edge with major knicks can be filed down, or ground on an **emery wheel**, but be careful not to heat the steel in the process, as you'll weaken the metal by negating the steel-hardening heat treatment that was imparted during manufacturing. File the same amount on each side to keep the edge centered.

Normal wear and tear can be addressed with coarse and fine stones. Your Japanese water stones can be used on your hatchet, but palm-size **ax-sharpening pucks** are better suited for the convex bevel of ax blades.

These round stones generally have a course side and a fine side. Using water, work the coarse side of the stone in a circular motion along the edge of the hatchet. Keep to the original bevel, which is likely about 25° to 30°, and if you're having trouble telling what angle to hold the stone, use that permanent marker trick to be sure you're sharpening all the way to the very edge. Keep at it until you've formed a burr along the edge, then give a few light swipes along the length of the edge with the stone to remove it. Repeat this with the fine-grit side of the stone.

Stropping will put a hair-splitting edge on your blade, and this is imperative if your hatchet is used for wood carving. It's less necessary for axes used to split wood, as these work better splitting the grain than cutting into it.

Knives

Knives have been in constant use for millions of years, even pre-dating *Homo sapiens*—cutting cord, preparing food, trimming hair, and waging war. Even if most don't realize it, this tool has always been, and remains, a critical implement in everyday life. If there's any doubt, bear in mind that even the most self-proclaimed of tool novices have surely prepared a meal with the knives in their kitchen.

So yes, all knives are tools, including those found in the dishwasher. However, the focus here will be on knives found on the jobsite, in the toolbox, or in the shop.

First, let's consider the **pocketknife**. It's a concept more than it is a specific tool: a small personal knife that's kept handy. Normally, this means a folding knife tucked into a pocket, but a modest fixed-blade knife in a sheath could also be called a pocketknife without much debate.

Sharp pocketknives are indispensable tools that require little in the way of bells and whistles. However, don't be fooled into thinking that a cheap knife is acceptable. When you lock the blade open and grasp the handle, your fingers are now in the path of the closing blade. Would you trust your favorite pinkie to a gas station bargain-bin knife? Don't skimp on locking knives.

You'll want to keep your knife sharp, and for exactly how, see the sidebar on sharpening (page 93). It can take some work to keep the blade of a knife sharp, and if you think the effort is not worth it, especially for a pocketknife that might see a lot of abuse and wear, you're not alone. For this reason, we have **utility knives** with their disposable razor blades, perfect for everyday tasks, such as opening boxes, stripping packaging off pallets, or cutting electrical wire (a blade-ruining job).

> The simple design of a pocketknife hasn't demanded much evolution. The oldest known pocketknife was dug up in Austria and dated back to 500 BCE. It had a bronze blade, a bone handle, and a familiar shape. Roman knives from the first century look remarkably modern as well. These folding knives had no lock—a style of folding knife called a **friction folder**—and it's a design that's still popular in countries that limit the carrying of locking knives. European **peasant knives** from several hundred years ago, with their round wooden handle and locking collar, look exactly like modern counterparts still produced in France by Opinel. The Japanese **higonokami** pocketknife hasn't changed since the 1800s, and another similarly simple design, the French **douk-douk**, is still being made a hundred years on.

Other niche knives can be found around the shop, including hooked-blade knives like **linoleum knives** and **carpet knives,** and **marking knives,** which are popular in Japan. The stout chisel blades of the **kiridashi** are excellent at scoring a line across the grain of lumber. Wood carving, too, uses its own battery of specialty knives, like the **sloyd** or the **chip-carving knife**. And to excavate the depressions in wooden bowls or spoons, there are curved-blade carving knives, such as **hook knives**, **scorps**, and **inshaves**. Sharpening these curves can be a real challenge, so it's more important than with your average knife to keep their blades away from anything harder than wood and to frequently strop their edges.

UTILITY KNIFE

HOOK KNIFE

CARPET KNIFE

SLOYD CARVING KNIVES

SLOYD WITH BOLSTERS

FRICTION FOLDER

HIGONOKAMI

KIRIDASHI MARKING KNIFE

Oscillating Multi-tools

You'd be excused for believing that the **oscillating multi-tool** is a new invention. In truth, it has existed since 1967, when the German manufacturer Fein introduced the power tool as a medical device for removing casts. The tool was purpose-built for cutting away the hard pack of plaster and gauze while sparing flesh.

Fein recognized that the multi-tool might have value beyond the doctor's office and marketed it as an automotive tool in the 1980s, touting its ability to cleanly remove windows that had been fixed in place with silicone. But it wasn't until Fein's patent expired in 2009 that these tools exploded onto the cordless tool scene. Now, hardly a home renovation is performed without one.

The advantage of these tools, aside from their assortment of blades and attachments, is that they do not rotate or reciprocate a blade. They work by vibrating an adjustable blade back and forth, and only just so. This minor difference in action makes all the difference in access, giving them the ability to cut in ways that no other tool can.

A multi-tool can cut away the flooring around trim, use a scraping blade to peel up old adhesives, and then slice away the bottom of existing baseboards to accept new flooring.

Much like the reciprocating saw and the angle grinder, there are attachments for everything. You can buy blades for cutting metal or tile, removing grout and caulk, and cutting wood where you might encounter hidden nails. Triangular sanding pads, which can get into corners better than many other detail sanders, are also available. The fine teeth of the metal and wood blades, combined with the high-speed oscillations, are quite capable of cutting plastics, though there are also specific blades for those materials. There's even an attachment that turns the tool into a miniature scroll saw. If you need to remove a piece of drywall—a common task in home repair—the thin blade of a multi-tool produces less dust than most. And if you do need to cut an outlet hole in your drywall, you can buy a rectangular box blade that cuts the entire hole all at once!

But for all the multi-tool's capabilities, there are some considerations. Most importantly, as an uber-capable multi-tool, it does a lot of jobs, but it doesn't do any of them with perfect precision. For some jobs, it works remarkably well, like when you need to lay the saw blade flush on a surface that will act as a guide. But in most applications where the tool is chosen because of its unique ability to access a spot and plunge into it, the cuts can be a little rough simply because it's handheld and doesn't have the ability to be steadied against a surface. Just as with saws (see page 110), you're trading accuracy for dexterity.

Pipe and Tube Cutters

There are a lot of different types of pipes and tubes and an array of tools that will cut them, though the right tool for the job will make for an easier, cleaner cut. Here's how pipe and tube cutters shake out.

PVC, ABS, polyethylene pipe (usually shortened to "poly pipe"), and PEX are plastic pipes used in plumbing and irrigation. Most sizes are best cut with a **ratcheting pipe cutter**. These cutters are opened wide and are placed around the pipe, then the handles are closed until they click. The handle is released and squeezed again. *Click*. Release and squeeze. *Click*. On it goes, the blade working its way down, until the pipe is cut in two. Larger PVC and ABS pipe might benefit from a hacksaw or reciprocating saw, while small poly pipe and PEX can be cut with a utility knife. But in general, all of these pipes are more easily cut with a handheld ratcheting cutter.

Flexible tubes (like the vinyl tubes used in a home brewing kit) are best cut with **flexible tube cutters**. Some models look like squat staplers with a razor blade tooth where a staple should eject. Others resemble a ratcheting pipe cutter.

Metal pipe, like the kind you'd use for indoor plumbing, should be cut with metal-specific **tubing cutters** (also called **pipe cutters**). Some tubing cutters are meant only for copper, but there are

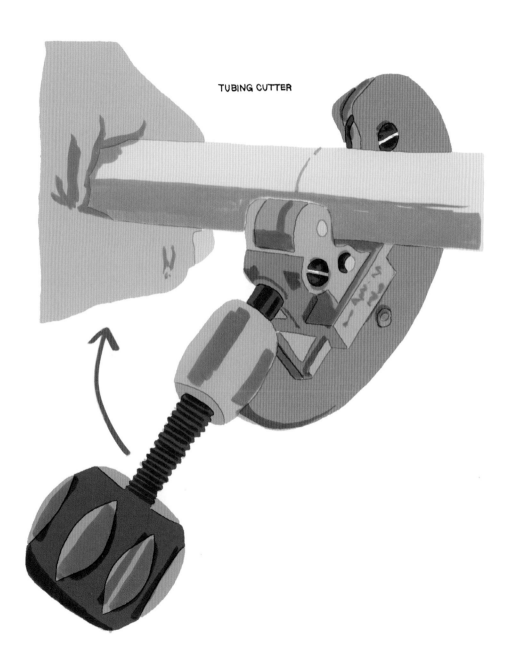

TUBING CUTTER

models suitable for multiple materials, including steel. These are useful for cutting conduit and other steel pipe and will do so more cleanly than most other tools. Tubing cutters are palm-size or a little bigger. They have a couple of guide rollers and a sharp V-shaped wheel of hardened steel or carbide, like you'd find on a glass cutter. The V-shaped wheel is attached to a threaded knob, which allows the cutter to be adjusted to the size of the pipe being cut. This knob also provides the constant application of force that eventually cleaves the pipe.

To use a tube cutter, open the cutting wheel enough to fit the pipe into the cutter and tighten down the blade until it is snug. Then rotate the cutter around the pipe. It will be hard to spin at first, but after a rotation it will get easier because the wheel has carved a groove around the pipe. Tighten the knob further, and rotate it around once again. Repeat until the pipe is cut through. It will be a clean cut, but a deburring tool should still be used on the interior of the tube, and for this reason many cutters have one built in.

Like any metal, a pipe *can* be cut with a saw or an angle grinder, just as flexible tubes *can* be cut with a utility knife, but these cuts will usually be rough or uneven, and if performed repetitively, tiring. A pipe or tubing cutter designed for the material you're trying to cut will make a cleaner cut that requires less effort and produces a tighter connection. Avoid the frustration and embarrassment of a hidden leak by using the right tool for the job.

Plasma Cutters

As the fourth state of matter, plasma gets the least attention, mostly because folks don't knowingly interact with it in everyday life as they do liquids, solids, and gases. But plasma is all around us. It is the glow within neon signs and fluorescent lights. The aurora borealis and aurora australis (those colorful dances of effervescent light visible in the night skies of the northern and southern poles) are also plasma. Lightning is plasma, as is the sun and all of the stars. In other words, though plasma is a comparatively less common state of matter here on Earth, it's by far the most common across the universe.

That star power should give some hint as to why the plasma born from the compressed air and electricity of **plasma cutters** is so capable of cutting metal. It's the heat! That concentrated jet emanating from the tip of a cutter's torch can be upward of 45,000°F, or about 25,000°C. That's four times hotter than the molten core of the Earth. As you'd imagine, that type of heat can do

Before using a plasma cutter, you'll want to don welding attire to protect yourself from sparks: a head covering, a flame-retardant shirt or a welding jacket, and gloves. If you're working at lower power, a full **welding helmet** might not be necessary and **#5 welding glasses** will suffice, but as power increases, so does the need for more protective eye shielding.

some damage, and that's also why it's such an effective metal-cutting tool.

Unlike an **oxyacetylene torch**, which can only cut iron-containing metals, plasma cutters are capable of cutting *any* type of metal. However, you sacrifice portability for this increase in capability. Plasma cutters require power and often must be connected to an air compressor. Some models do have a small, built-in air compressor that allows them to cut thinner materials as a stand-alone unit, but they still require an external air hookup to cut thicker stock.

A note about the incoming air: It's imperative that the air coming from a compressor be dry. Use an **external air dryer** to collect compressor condensation if the plasma cutter doesn't have one built in. Moisture in the incoming air will cause the torch's electrode to burn up, producing poor cuts.

The thickness of the metal being cut will dictate the size of the machine required, whether it demands external air, and the appropriate power setting (in amps). On many cutters, switching between the power ranges will also demand swapping out the size of your consumables in order to match the power output. These deteriorating parts—the consumables—are the nozzle, the electrode, and a few other inserts that wear over time.

Before any cutting can begin, the workpiece clamp and cable need to be attached to complete the electrical circuit between the torch and the material. Fasten the clamp to the main workpiece, not to a part destined to fall away as a cut is made.

Cutting requires a steady hand. Find a stable position and practice the cut before squeezing the trigger. The tip needs to be kept ⅛ inch [3 mm] off the surface. Dragging the tip on the surface will erode the consumables faster unless your plasma cutter is fitted with a **drag shield**—an extra tip made for protecting the nozzle.

Moving at the correct speed is crucial for producing clean cuts. There is no set rate; it varies based on the type of metal, its thickness, and your power. But you'll know if you're moving too fast, as sparks will kick up at you. Conversely, if you're moving too slow, the sparks and molten metal will blow through the workpiece and straight down. This slow speed will also carve a wider and sloppier *kerf* (the channel left by the torch as it removes material, not unlike a saw kerf). You'll know you're moving the torch tip at the correct speed if the sparks are sent out through the underside of the workpiece but kick back at a 10° to 15° angle, trailing behind the torch.

A skilled hand—especially with the aid of a clamped guide or template for the tip to trace—can slice cuts that significantly minimize cleanup work. Even so, the best cuts will require some tending to. The *dross*, the scraggy bits of hardened waste that form on the underside of plasma-cut metal, can be removed with a **file** or a **chip hammer**. Once all that's done and before the plasma cutter is stowed, be sure to take note of the condition of the consumables, as their eventual wear will produce sloppier cuts and even more dross to contend with.

Saws

One of the first tools humans ever created was a piece of stone chipped to form a cutting edge—a process known as *knapping*. Those stone blades were used to saw through bone and wood, but their varying degrees of serration muddy the line between what we'd label a saw or a knife. It wasn't until a few thousand years ago, when metal replaced stone, that saws came into their own and began morphing into fine-tuned specialty tools.

One way to think about saws is that when you choose a saw you can optimize for dexterity or accuracy, but generally not for both. If you want a straight line at a perfect angle across an irregular object, you can choose a saw that will be rigid and limited in its movement. Conversely, if you want the mobility to cut a Florida-shaped hole in a piece of plywood, you can use a maneuverable saw with a more flexible blade, but any attempt at straight lines won't be so easy. So there are trade-offs.

Hand Saws

The basic **hand saw**—that utilitarian stalwart hanging in your parents' garage and your grandparents' garage before that—is probably what you think of when you think of a saw. The design dates back nearly five thousand years to ancient Egypt, where copper saws were used to cut wood and stone. Modern saw blades look very similar to early iterations, but there have been a lot of

CROSSCUT SAW

JAPANESE PULL SAW

MITER SAW

BOW SAW

HACKSAW

COPING SAW

RECIPROCATING SAW

JIGSAW

PORTABLE BAND SAW

CIRCULAR SAW

POWERED MITER SAW

DRYWALL SAW

advancements: improved metals, thinning the back of the blade to reduce binding, teeth patterns and angles optimized for different types of cuts.

These commonly encountered saws are typically **crosscut saws**, meaning that they're best used to cut a piece of wood to length across its grain. Crosscut teeth are tightly packed and sit at a steep angle, or *rake*. They can be used to *rip-saw*—that is to cut a piece of wood along the length of its grain—but it will be such slow going that you'll soon wonder if the ancient Egyptians had any better ideas. (They didn't. The first **rip saw**—utilizing fewer teeth and a shallower rake—wasn't invented until the fifteenth century, and they're still available for purchase, if a bit less common.)

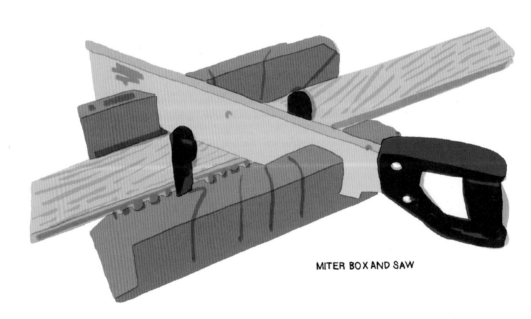

MITER BOX AND SAW

These hand saws cut on the push stroke only, which is helpful for powering through dense wood, but the blades have a tendency to buckle under the force, making for an imprecise cut. A special category of hand saws, called **backsaws**, address this problem with a metal rib fixed along the spine that stiffens the saw blade. They're meant for delicate work and are generally named for their specific task. For example, hand-powered **miter saws** (often paired with a **miter box** to guide the blade) are used to cut the precise angles where wood meets in a corner, known as a *miter joint*. There are also **tenon saws**, **dovetail saws**, and **sash saws**, which are used to make cuts for other tight-fitting wood joinery.

Japanese Pull Saws

These saws got their start in the toolboxes of Japanese carpenters but are now found in most woodworking shops around the world. Many of these pull saws feature a crosscut tooth pattern on one side of the saw blade and a ripping pattern on the opposite, making them more versatile than western crosscut saws. They have a long straight handle with enough room for two hands, which can make for a more consistent stroke. The thin blade produces a narrow *kerf* (the channel created by the saw blade as it removes material) and cuts on the pull motion only. This tension, made by pulling, keeps the blade stiff and allows for a more measured cut, even by a novice.

The Japanese pull saw's less powerful stroke and delicate construction come from its original purpose of cutting softwoods common to Japan, but these days they're regularly used on the hardwoods found in furniture making and other fine woodworking.

For tougher tasks, lubricant can be sprayed on the blade to ease it through dense, or sap-rich, wood. They're indispensable wood-working tools and, given the availability of models with folding handles, can be conveniently stashed in any tool bag.

Hacksaws

The **hacksaw** is another staple of each generation's basic toolbox. Even the most novice DIY homeowner has had to cut something other than wood and wondered if their inherited **hacksaw** was the tool for the job. Hacksaws have been the go-to tool for many a handyperson doing plumbing, removing an old padlock, or cutting through a rusted bolt.

Hacksaw blades, which are mounted between the arms of the saw's C-shaped frame, are replaceable and come in a variety of tooth patterns. The teeth are smaller and more numerous than those found on most wood saws, making them best suited for cutting metals and plastics. Most blades have an arrow printed on them that advises pointing the teeth forward, so that the saw cuts on the push stroke. The thin blade is semi-maneuverable, but hack-saws aren't great for cutting curves with any precision.

When you're armed with a basic hand saw and a hacksaw strung with a general-purpose 18-TPI blade, there's hardly a common household object that you can't cut to size.

Coping Saws

Coping saws are a type of **bow saw**—an arched frame that holds a blade in tension, much like a bow string. Wood-framed bow saws are one of the oldest saw designs and are still used today by traditionalists. Once the bow design is scaled down and the blade made skinnier, you end up with a coping saw.

Coping saws are meant for cutting curves and other shapes in wood. Blades for cutting thick metals do exist, but they are less common. (Similar-looking saws, known as **piercing saws** or **jeweler's saws**, are frequently used in fine metalworking.)

The blades of coping saws are detachable so that they can be replaced but also so that the saw can be reassembled for interior cuts with the blade passing through a hole in the workpiece. The handle is parallel with the blade, pointing back at the user, which limits the amount of power that can be put into the stroke. Coping saws make delicate and measured cuts on the pull stroke only, and the handle pivots, letting the user rotate the saw blade to cut those curves. This allows the frame to be rotated independently of the

It's not just about choosing the right saw for the job. If your saw has replaceable blades, it's also important to use the correct blade for the piece being cut. Blades with more teeth are used for thinner or harder materials, while blades with fewer teeth are used when the workpiece is thicker and softer. In the United States (and, curiously, most other metric-system countries) the density of saw blade teeth is measured in TPI (teeth per inch). Most blade packages and saws will specify both the TPI and what materials the blade is suited for, but these charts can also be looked up easily.

saw blade, which means the frame can be turned sideways and out of the way for a long cut down the length of a workpiece.

If you desire a powered coping saw, a **scroll saw** is what you're looking for. These benchtop units reciprocate a thin saw blade up and down. The workpiece is fed through the blade the way you would feed fabric through a **sewing machine**. That isn't the only similarity they share: Both early sewing machines and early scroll saws were powered via a *treadle*—a foot pedal that spins a flywheel. These days they're electric, though some scroll saws still use a foot pedal as a control switch.

Ultimately, whether it's a scroll saw, jeweler's saw, coping saw, or some similarly shaped saw yet encountered, the principles are universal. Delicate saw blades, *held tight on both ends*, are ideal for cutting accurate curves.

Jigsaws

A **jigsaw** is a power tool used for cutting curves where more speed is desired and less accuracy is required. It's held with one hand and has a flexible blade protruding out the bottom. Because there is no frame around the saw blade (as there is on a coping saw), it can be plunged through a hole in the workpiece and maneuvered unrestricted, cutting any manner of convoluted shape. Your jigsaw puzzles weren't actually cut with a jigsaw (in fact they're stamped out), but you can see where they get their name.

The blades are replaceable, and like other disposable blades, they come in different lengths and a variety of tooth styles made for wood or metal with various TPI counts (see page 115). Bear in mind the trade-off: A blade with more teeth and a finer cut will also be slower going.

When cutting wood, take your turns slowly, as the blade can easily bend and cuts won't be plumb. Reducing your speed will also produce cleaner cuts with less splintering, but masking tape placed along the line you intend to cut can also help prevent this tearout. Another trick is to use a marking knife to score a line before running the saw blade along it. Or you can simply cut the wood upside down, as the cut on the underside where the blade pulls in will be cleaner than the top face where the blade exits. For improved sheet metal cuts, sandwich the metal between two layers of plywood. And when cutting thicker metals and steel, use plenty of cutting oil to keep things from getting too hot.

Jigsaws are exclusively power tools, but if you're looking for a hand-powered tool of similar utility, you might consider **keyhole saws**. In addition, there are hacksaws that hold replaceable blades by only one side, allowing them to be inserted into holes for cutting curves in plastics and thin metals.

Reciprocating Saws

Reciprocating saws are power tools best left to rough work that requires brutality and little in the way of patience. It's often said "They're the right tool for the job when you don't have the right tool for the job." These saws offer a variety of disposable blades made for pruning trees, cutting metal, and even dicing up nail-ridden lumber. When Milwaukee Tool invented this saw in 1951, they branded it the **Sawzall**, as in, it *saws ALL*. It's the tool of choice for demolishing an old shed or cutting the roof off your Chevy Blazer and then (speaking from experience) wishing you hadn't.

Reciprocating saws are nimble tools. The saw is long and pointed, with a thin, flexible blade poking out the front, like a sword, which is why they're sometimes called **saber saws**. Reciprocating saws are perfect for cutting in a tight space, but the nature of their dexterity and adaptability leaves them inherently unstable.

There are a few ways to tame the saw's rowdy leanings. For starters, select the correct type of blade and the shortest one that'll get the job done. Know what you're cutting and what's behind it. The *shoe*—a flat metal pad found at the base of the blade—should be adjusted to further keep the blade protrusion as short as necessary and held firmly against the surface of whatever is being cut, otherwise the saw will jump around.

The trigger on a reciprocating saw allows you to control the speed of the reciprocation, so start slow. Work a groove and build up to a reasonable rhythm, all while holding the saw blade and shoe firmly against what is being cut. Use cutting oil when using hacksaw-type blades on thick metal. And change blades when they're worn or damaged, as you should with any tool that uses disposable blades.

Band Saws

Band saws are generally encountered in two forms: floor-standing shop fixtures and handheld models about the size of a briefcase. Both power tools spin a continuous ribbon of blade around the whole machine, offering a relatively narrow opening where materials can be fed into the moving teeth. It's a design so efficient that virtually all commercial lumber today is milled via band saw.

The long blade of a band saw means less wear on each tooth, more heat dissipation, and less metal fatigue. The allure of that longevity was apparent early on, and the technology was patented in 1808 before usable blades for it actually existed. Forty years later, Anne Paulin Crepin invented a welding technique that finally allowed the blades to withstand the constant flexing, and the modern band saw was born.

Portable band saws look unwieldy and off-puttingly specialized, but they're surprisingly useful and can be indispensable in making quick work of metal, plastic, or even wood. That said, the nature of having to heft the band saw and maneuver it unsupported through a workpiece can cause a cut to wander. You'll want to plan a cut you can remain steady through while maneuvering the saw. When cutting metal, use plenty of cutting fluid to lubricate the operation and ensure a cleaner cut.

Floor-standing band saws tend to the previous issues by offering a flat table on which to set the workpiece and a blade held firmly in place on both ends, not unlike a scroll saw. Two-handed control and perfectly perpendicular cuts make it a go-to tool in any shop that's got one. Aside from being found in fabrication shops, floor-standing band saws also happen to be butcher-shop work-horses, happily spending hours reducing big meat to little meat . . . so watch your fingers.

Circular Saws

Circular saws are rightfully among the first power tools that many people buy. Anyone who has tried to cut a volume of wood with a hand saw will quickly wonder if there's a more efficient and less backbreaking solution and will either find their way to a circular saw or to the couch as they throw in the towel. Circular saws are powerful, portable, and nearly as nimble as a hand saw.

The advantage of these tools over traditional hand saws (aside from speed) is their precision. Circular saw blades protrude through an adjustable shoe that sits flat on the surface of the workpiece, so the angle of the cut is always consistent. This angle is most often square, which is to say 90° straight up and down from the face of the workpiece. But the angle can also be adjusted on most saws to at least 45°. So, the saw keeps the angle of the blade consistent, but what about cutting a straight line? That onus still falls on the user, and this requires a sharp eye and a steady hand, but various jigs and tracks can help.

Most saws include a **rip guide** for making rip cuts up to several inches wide. Some pricier saws are sold as dedicated **track saws** that ride along a section of track atop the workpiece, cutting dead straight. But clamping a piece of wood down as a guide works nearly as well.

The saw blade cuts upward, which leaves a tidier surface on the underside of the material (where the blade pushes in) and a splintered face on the top (where the blade exits). So if appearances matter, cut with the visible face down. Alternatively, you can score a line with a knife or use removable masking tape along the face of the cut to prevent splintering.

Circular saws offer a particular danger in that they are often used one-handed (while the other hand is steadying the workpiece), and they can buck or kick back, which can leave the exposed hand vulnerable to grave misfortune. If you make a mistake here, you'll be able to count on one hand how many good fingers you own.

Here are the basics of how to use a circular saw. Before doing anything that requires messing with the blade—and this goes for any power saw—be certain the saw is disconnected from power. Install the correct blade. Set your cut depth by adjusting the shoe so that the amount of blade protruding isn't any longer than necessary. Use a square to check that the angle of the blade is correct. Then connect the power. Before beginning a cut, make sure your hands are placed out of harm's way. Don't start a circular saw—or any saw—with the blade touching the material. Instead line up the cut with the saw blade a small distance away, then squeeze the trigger (there may also be a safety you need to simultaneously disengage with your thumb), and push the saw into the workpiece.

The piece being cut should be supported along its length so that it will remain supported throughout the whole process, even once it's cut in half. Usually this is done with four or more support blocks, but one trick that's helpful with large sheet goods, such as plywood, is to support it on a full-size disposable cut mat of Styrofoam insulation.

Any twist can pinch the blade, and a binding on the rear of the blade can cause the saw to kick back dangerously. This can also happen when wood grains cause the kerf to pinch closed. In some cases, it may be necessary to wedge a long kerf open to safely continue the cut.

Most of these saws have a cut depth adequate for standard dimensional lumber. But if you're in need of more blade real estate, say for cutting large timber, look into **beam saws** and **beam cutters**. A beam saw looks like a circular saw, but more than double the size, while a beam cutter is a vertical chainsaw blade mounted to a circular saw body. It shouldn't go unsaid that their wood-devouring capabilities are impressive to behold.

Miter Saws

Powered miter saws are angle-cutting machines. They're capable of producing a variety of complicated cuts and will do so with great precision. The trade-off for this amount of functionality is diminished portability. Even the most compact versions will take up significant garage space compared to all the saws previously mentioned. Regardless, they're indispensable tools for creating quick, accurate cuts and for joining corners.

While larger and more intimidating than circular saws, miter saws are surprisingly user-friendly and generally safer for novices. Even without exploring the range of their capabilities, these saws work well as a rudimentary **chop saw** (a term for the miter saw station that's sometimes thrown around on a jobsite) for making quick square cuts. It's easy to set the material up along the saw's flat work surface and right-angle fence, which keeps things steady. Most miter saws have built-in clamps so hands can be moved away from the blade, and many have lasers to demarcate the cut line. There's also a cover that retracts only as the blade is pulled down and into the workpiece. Plus, guide rails keep the blade on a set path, and the entire apparatus is so confined in its range of motion that any risk of kickback or bucking is effectively removed, allowing you to confidently and safely make accurate cuts that are comparatively free of hazard.

Angles can be adjusted in two directions to create compound angle cuts. Most saws have presets for common angles, but these should be checked with a **square** or **digital bevel gauge**. (You'd also do well to read about the peculiarities of miter angles in the discussion of miter protractors; see page 21.) Support blocks should be used on workpieces that are longer than the saw's base, and some miter saw stands have extensions for this purpose.

A significant concern is that miter saws are professional tools with a professional degree of freedom in how they can be manipulated, to the extent that the saw can be set to an angle where it will cut into itself. It's imperative that you move the stationary blade through the full range of the cut during setup to ensure you're not about to cut into the frame of the saw during operation.

Table Saws

The **table saw** is the only saw thus far discussed that is designed for making long ripping cuts of a consistent width along the length of a piece of material. They're generally intended only for wood, though blades can be installed for plastics and some metals. If you want to rip a piece of wood down to an exact width across its entire length, there's no replacement for a table saw. Likewise, because the blade can be set at an angle or to any depth, this is the ideal tool for making bevels or for cutting channels (called *dado* cuts), such as when overlapping edges or nesting shelves.

Compact jobsite versions are approximately the size of a miter saw, and because they require a stable base, they are often transported on folding stands. This entire setup can fill half a truck bed and is only "compact" compared to the larger table saws found in many shops that can command a whole room.

Table saws are notoriously dangerous, both in their tendency to swiftly remove wayward digits (ten fingers per day in the United States!) and in their ability to throw wood. (A saw can hurl an improperly handled section of lumber across the room and into a wall.) Their danger is not to be taken lightly, which is why manufacturers include a variety of safety features.

Many saws include a *riving knife*, which is a shark fin of metal behind the blade. This prevents the kerf from closing behind the cut, which would otherwise pinch the back of the blade, causing the material that's being cut to rise, kick back, and ultimately be flung at the user. Some saws also have serrated cams, known as *pawls*, which allow material to be fed through but prevent the material from rising and will lock down to prevent kickback. Additionally, many table saws have a safety cover that caps the top of the blade.

But the biggest development in table saw safety came from a company called SawStop. In 2004, they began selling saws that can detect the capacitance of human skin on the saw blade while it's running. (Your cell phone's touch screen also detects capacitance, which is the measure of how well a material holds an electrical charge. This is how it can tell the difference between your finger and other objects.) On these saws, as soon as skin is detected, a perforated aluminum block is sprung into the blade, jamming it and hopefully preventing injury. These replaceable cartridges (and a damaged blade) aren't inexpensive, and the saw's safety mechanism can be falsely triggered by wet wood or metal, such as nails. Still, it's less costly than a severed digit.

Here's a general description of how to use a table saw: First, install an appropriate blade. The blade protrudes through the table, like an inverted circular saw, so raise or lower it via the hand crank to a height that is sufficient for the cut, but not much more. Then adjust the guide fence the appropriate distance from the blade.

Keep in mind you'll want to try to set up your material so that the waste—that is, the part being trimmed off—is on the outside of the blade. This is because any waste caught between the blade and the fence is liable to be shot back toward you at high speed. Then turn the saw on and push the material steadily and evenly against the fence, through the blade. As your hand nears the blade, you'll want to transition to pushing the material with a **push stick.** Once you've pushed all the way through, power the saw down. Do not reach in to

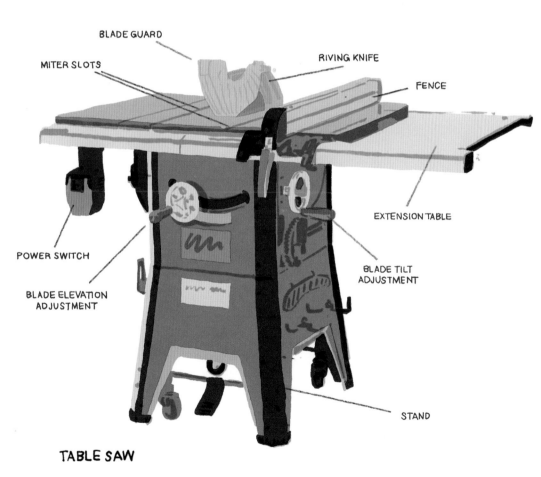

BLADE GUARD

MITER SLOTS

RIVING KNIFE

FENCE

POWER SWITCH

EXTENSION TABLE

BLADE TILT
ADJUSTMENT

BLADE ELEVATION
ADJUSTMENT

STAND

TABLE SAW

remove any items or clean any debris until the blade has completely stopped rotating.

Regarding that push stick: It's important to use a scrap piece of wood to hold the material flat and against the fence for the final push of a cut, where your hand would otherwise be too close to the blade. But there are also adjustable plastic **push blocks** that can clamp over wood and put a solid barrier between the blade and your hand; they are particularly useful on narrow materials. A **featherboard** can also help keep the workpiece tight against the fence.

Angled cuts can be made by using a **miter gauge** (not to be confused with a miter protractor) that fits into the saw's miter slots. It's also possible to crosscut with a table saw, but the saw's accuracy and safety are predicated on the smooth and straight movement of a workpiece through its large-diameter blade, so you'll need a **crosscut sled** to support the workpiece and make this possible. They can be purchased or fabricated, but either way they're critical to the process. Attempting to crosscut freehanded will inevitably be fraught with complications and may cause injury.

Tile Saws

Tile saws are wet saws, meaning that water flows over the blade and the tile that's being cut, cooling the blade and the material and assisting the diamond-coated blade's abrasion through the tile. They can be used to cut stone tiles, such as marble and granite; porcelain tiles; and even glass. Tile saw blades are rough to the touch and might be notched, but they don't have the familiar teeth seen on most saw blades. That slow and well-lubricated grinding away of material is intrinsic to all stone cutting, and similar tools can be found cutting large granite slabs or shaping precious gems. You may even see a **gas-powered wet saw** being used to remove a chunk of freeway concrete.

Wet tile saws can be used in place of, or in conjunction with, other tile-cutting tools. Plier-shaped **nippers** are good if you're just making a few quick cuts. **Tile cutters**, which look similar to a paper cutter, work by scoring the tile before snapping it along the line. **Diamond blades** are available for angle grinders and circular saws, but the lack of lubrication (except on a few specialty **wet circular saws**) means that they should be used for short periods and sparingly. (A smart application for angle grinders is in completing electrical outlet holes where the full-size tile saw blade is too large.) Lastly, for maximum intimidation, there are **wet chainsaws** that will slice through stone, brick, tile, and all the concrete beneath. They're wildly impractical for the average user, but it's satisfying to watch videos of their well-lubricated mayhem.

All wet tile saw parts and blades should be rinsed off and dried to prevent rust. Sediment will build up, so replace the basin water before it starts to clog the pump inlet. Alternatively, many saws can be configured to use a supply of fresh water rather than circulating runoff.

Snips

There are a few varieties of **snips**, but they're all used for cutting sheet metal, like what is commonly found in HVAC (heating and air-conditioning) ducts, flashing, gutters, and some auto body parts. The first thing to know about cutting sheet metal is that its thickness is measured by its gauge, and all snips have a maximum gauge they are capable of cutting. Confusingly, the bigger the gauge number, the thinner the stock. The reason for this confounding system is that metal gauge measurements descend from wire manufacturing, and a wire's gauge indicated how many times it had to be *drawn*—that is, stretched out through a **tapering die**—to be made thinner. For example, a 2-gauge wire was drawn through a tapering die twice, while a 20-gauge wire was drawn twenty times and made all the thinner in the process. Thus, the bigger the number, the thinner the metal.

Tin snips look like overbuilt scissors and are generally capable of making straight cuts in 22-gauge cold-rolled steel or stainless steel sheet metal that's a little thinner. The exact gauge that tin snips can handle varies depending on the material being cut and the model of the snips.

Aviation snips are more commonly used because of their ability to cut thicker sheet metal—18-gauge steel or 16-gauge aluminum—but they are also the most commonly misunderstood of the snips. They come in three colors: red, green, and yellow. Most folks think these are left- and right-handed distinctions, while some think it's just so they can match their snips to their shoes—but this is not the case. The burly construction of aviation snips, and the thick metal they're cutting, means they can't easily move through metal like a pair of scissors through paper, so each color is designed to navigate in a specific way.

Red snips are meant to be used on the left side of a workpiece, with the waste on the left and the piece being trimmed on the right. The waste will curl up and away, and the workpiece on the right will remain flat and in pristine condition. These also work well for cutting curves to the left. Green snips are for cuts with the waste to the right and for curves to the right. Attempting to work in the opposite direction the snips are designed for will mangle the workpiece. Finally, yellow snips are for short straight cuts down the middle of a workpiece.

Those left and right colors on aviation snips aren't arbitrary designations. They correspond with the navigation lights on aircraft—a red light on the left wingtip and a green light on the right. Those lights exist so you can tell if a plane is flying away from you or headed toward you on a collision course. Ships (and even some spacecraft!) use this same pattern of navigation lights.

All handheld snips will have some difficulty making a clean cut through the middle of a large section of metal. They work best taking short nibbles and trimming off a small section of waste that can curl out of the way. Cutting through the center of a gutter, for example, will likely leave both sides of the cut a little worse for wear. For cleaner cuts, first make that rough pass, then trim an additional pinkie's width of metal off with the proper snips, to get a more finished edge.

For larger jobs where your grip strength isn't up to the task, like cutting multiple panels of roofing, there are electric **nibblers** and **shears**. These can handle the undulations of corrugated metal with ease, can cut curves, and will quickly chew through a project. They're available both as dedicated tools and also as drill attachments.

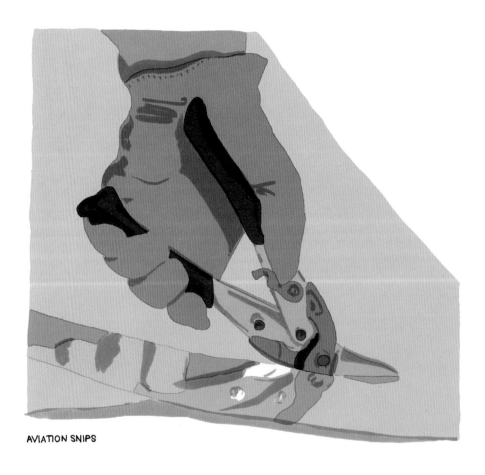

AVIATION SNIPS

Wire Cutters

Wire cutters are built into quite a few tools. Many **wire strippers** and pliers have wire cutters. And what are bolt cutters and cable cutters if not large-diameter wire cutters? Although the definition can get a bit nebulous, generally speaking, the term wire cutter is reserved for **diagonal cutters**, which are one-handed plier-like tools that feature stout bird's-beak jaws capable of cutting small to medium-size wire.

Depending on where you are in the world, these cutters go by a few other names: *snips* or *nippers* in some European countries, **side-cutters** in and around Australia, and **dikes** in the United States (presumably a lazy approximation of *diagonal cutters*).

Wire cutters are a common electricians' tool. The pointed jaws are better suited for reaching into tight spots than the blades that are buried in the rear jaws of other pliers, plus they can be used to strip wire if they're sharp. Basic diagonal cutters can also be used for snipping barbed wire and other similar farm material, but if heavy-gauge wire, nails, or a long line of fence severing is in your future, you'd do well to upgrade to a more powerful pair. Go with a well-regarded brand and opt for longer arms. Better yet, Knipex manufactures a set of **high-leverage diagonal cutters** that use a double fulcrum to produce twice the force with half the effort.

You might see diagonal cutters used in low-voltage electrical work, but more commonly this would be a job for their smaller cousin, **diagonal flush-cut pliers**. These are better suited to the small-diameter wires typically encountered when working with electronics, as well as the work of clipping the leads on soldered components. Nipping these leads with the recessed bevel of larger diagonal cutters leaves a lengthy and sharp remnant, but flush-cut pliers avoid this by cutting, well, flush. They're capable little tools and generally sharper than their larger counterparts. Just be sure you don't test the limits of their delicate jaws on wire that's more than they can handle, as you'll quickly lose that sharp action that makes them so nimble.

Wire Strippers and Crimpers

Over the years, electrical wires have been encased in wax, oil, paper, rubber, and cloth. This sheathing insulated the bare wires from each other and other points of contact, so these coatings became known as the wire's *insulation*. In the mid-twentieth century, the wire industry pivoted toward plastics, and now practically all wire is insulated with PVC, vinyl, or similar synthetics. To work with these wires—be it soldering them, crimping them together, or screwing them to a light switch—some of that plastic insulation needs to be stripped away, exposing the bare wire.

People use a lot of terrible methods to remove insulation from a piece of wire: melting it off with a lighter, for example, or using a knife or—more disturbingly—teeth. These might seem tempting in their accessibility (your teeth *are* conveniently stored right there in your face), but a better tool for the job is a **wire stripper**.

Wire strippers come in two forms. The first is a basic pair that looks like a flattened pair of pliers. These have sharpened notches of various sizes to accommodate a range of wire gauges. To use a pair of wire strippers, place the wire in the appropriate notch, squeeze the blades closed, cutting through that outer layer, and slide the plastic insulation off. Most models have wire cutters built in, as well as **wire crimpers** and a knurled tip that acts as a small set of pliers. Strippers with a spring to return the handle to an open position will save on some hand fatigue during repetitive work.

The second type of wire stripper is the **automatic wire stripper**. These come in a couple of different forms, but the basic idea is the same. With a squeeze of the grip, these grasp the wire, sever the insulation, and slide it off—all in one motion. Aside from it being a simpler process, automatic strippers can also prove invaluable in situations where manual strippers cannot be used—for example, where the wire is barely protruding from some fixture. With manual strippers, you would have to pull the far side of the wire away with the opposite hand to slide the insulation off, while automatic strippers complete the task on their own.

Manual strippers, and some models of automatic strippers, require you to select the appropriate diameter die for your wire gauge. These work fine, but **self-adjusting automatic wire strippers** are an improvement. With these, wires of any commonly encountered gauge can be gripped and stripped, with no concern given to their exact size. These tools are sort of magical in their ability to self-adjust, and they are shockingly affordable. Many have adjustable backstops that can be set to create the same length of exposed wire every time, and like manual wire strippers they also tend to include wire cutters and wire crimpers.

Wire crimpers are used to compress the connectors that hold electrical wire connections together. These connections can be unique, so crimpers need to be matched to what they're crimping. Most common are the splices between wires and terminal connectors, like those that attach to a battery. These connectors are color coded by size and can be matched to most general-purpose crimpers. But then there are also specialized crimpers, such as **Ethernet crimping tools** and **coaxial cable crimpers**.

Like wire strippers, crimpers also come in two primary flavors. First, there's the manual type, which is no more complicated than a pair of pliers. These will crimp a variety of connections, and each size of connector is color coded to the section of the jaws where it fits. These tools might also have wire cutting and stripping blades. They work fine on many connections, but on larger connections it's difficult to generate enough leverage to really pinch down the connector. After a few crimps, you may wonder if there's a more usable tool.

There is, and it's the **ratcheting crimper**, which uses an extra fulcrum to get better leverage and ratchet down one step at a time, applying ever-greater force with each click and squeeze. Because they are more specialized, they may work with fewer connection sizes—and of course they'll cost marginally more—but your hands will thank you.

FASTEN

Bench Vises

Since the Renaissance, it's generally been agreed that there are a half dozen rudimentary tools that create mechanical advantage; they're known as the *six simple machines*. These are the **lever**, **wheel**, **pulley**, **inclined plane**, **wedge**, and **screw**.

There's some overlap here—all of these machines produce advantage by spreading work over a distance—but those last three are nearly identical in how they operate. A wedge is just two inclined planes working in tandem, and a screw is a helical inclined plane kept in constant forward movement by way of rotation (like a treadmill). If you put a tight pattern on those screw threads, all the work of rotating the screw is being distilled to a fairly short distance of travel, producing an impressive amount of force.

How much force? The jack that allows you to raise a two-ton car all by yourself to change a flat tire does so with a screw. And in the mid-1800s, the entire city of Chicago was raised about one story higher than where it began, and this was accomplished with thousands of screw jacks manned by relatively few workers. Screws are powerful.

If you break it down to its components, a bench vise is merely a set of sturdy cast-iron jaws mounted on screw threads, with a long lever on the screw. But from those simple pieces, the tool has tremendous clamping power. Vises have been around for a few hundred years, but it wasn't until stronger cast-iron models were in production in the early 1800s that the tool saw its full potential. For those who could afford the cost of all that metal, they became—and remain—shop staples. You'll find them in metal and wood shops, standing in as a second set of hands or grasping a workpiece stronger than another human ever could. They're mounted on the back of trucks that are outfitted for welding, plumbing, and other trades. They're features of some drill presses and workbenches and are even clamped onto some electronics repair stations.

If you're in the market for a vise, you may see them branded a few ways. **Machinist's vises** are generally made of higher-grade iron and have a more precise jaw alignment. More utilitarian models meant to suffer abuses are sold as **mechanic's vises**, **pipe vises**, or just plain **bench vises**. There are some blacksmith-specific vises, such as the **post vise**, but these are less common with each passing decade (as are blacksmiths.) There's also the **woodworking vise**, and these are often built into woodworking benches.

The metal jaws are replaceable and are usually textured for grip, though in the case of the woodworking vise, softer wood blocks are used in place of metal. Many general-purpose vises also have a curved section that's meant to grip round material, but a dedicated pipe vise is a better choice if significant pipe work is on the table. That said, pipe vises are limited in the diameter of pipe they can grasp; **chain vises** are a good choice for extra-wide pipes. These tools strap a section of chain around large-diameter pipes and clamp down with immense and distributed force.

The benefit of a vise is its sturdy integration into the workbench and its unparalleled clamping power, but there are some alternatives that satisfy similar needs to varying degrees. **Clamps** are an obvious solution; they can be used to pin down an object along the edge of a workbench. **Toggle clamps** affix anywhere on the surface of welding tables and woodworking benches and snap down in much the same way that locking pliers seize hold. You can also use a **bench dog** or a **holdfast**. Both slot into the dog holes of a workbench or anvil; they can be used to clamp down a workpiece or give it something to press against in combination with the built-in vise.

What's up with the names *bench dogs* and *dog holes*? It's an oddly common naming pattern. The teeth at the base of a **chainsaw bar** are also called *dogs*. The spikes used to staple two logs together are **log dogs**. And the spiked pole used to leverage logs into place is sometimes referred to as a **cant dog**. Presumably we owe this pattern to a dog's tenacious bite. Or maybe early woodworkers were just short on fresh ideas.

All of these dogs and clamps will hold a workpiece steady—some quite firmly. But often there is no replacement for the holding power of a 50-pound [22.7 kg] block of cast iron, securely bolted to a workbench, which is why the design has remained unchanged for nearly two hundred years.

Caulk Guns

Caulk guns are used for dispensing beads of viscous sealant or adhesive from disposable cardboard or plastic tubes. (*Bead* is the technical term for a line of caulk.) You can find these beads around the house: for example, the silicone around a bathtub, the paintable caulk on the edge of a baseboard, and the glue behind a brick facade. There are hundreds of products available in these tubes, but they all get dispensed the same way—from a caulk gun.

The caulk (or other sealant or adhesive product) in each tube is contained behind a tapered plastic applicator tip. Before the tube can be used, you'll need to cut off a bit of the tip at a 45° angle. The more you cut off, the wider the bead will be. The less you cut off, the finer (and slower) the application. A utility knife works for this, but anything above the cheapest model of caulk gun will have a cutter built into the handle. After the tip is cut, the caulk is still trapped inside. You need to poke a hole in the tube's foil seal, which is inside the plastic applicator, down at the base where it meets the tube. Perforate the seal a few times to ensure good flow. This can be done with a framing nail or with the foldout seal punch rod that's found on many caulk guns.

After you've cut the tube and punched the seal, you can load the tube into the caulk gun with the 45° cut facing down. When you squeeze the lever, the gun's plunger applies pressure to the base of the tube, forcing the caulk out the tip. Move the tip along a path to lay a bead down.

After a bead is laid, it's often pushed into the joint and smoothed out. This step is actually where the word *caulk* comes from—the Old Northern French *cauquer*, meaning "to press down." There are **caulk spatulas** for this, but many people prefer to simply use a wet finger. A little dish soap mixed with the water can help smooth the process.

Like most simple tools, you have more complex options to dive into. There are **battery-powered caulk guns**, **pneumatic caulk guns**, and **sausage guns**, which cut down on wasted packaging material (all those applicator tubes) by wringing the caulk from a thin tube that resembles a sausage casing.

The caulk world is vast, but all these pro tools and optimizations are beyond the average user's needs. Most folks just need a basic caulk gun for a small job here and there, and *almost* any model will do . . . just don't buy the cheapest piece of flimsy metal in the store. The biggest issue with those cheap guns is that they continue to ooze caulk after you're done squeezing. You have to press the release lever, or rotate the rod, to relieve pressure to stop the flow. Inevitably, this doesn't work too well. Improved **no-drip caulk guns** stop the flow as soon as you let go of the lever, and they are well worth the money. These mildly deluxe versions usually have better leverage, which will save a lot of hand ache, and their consistent pressure will elevate even the most remedial caulking game.

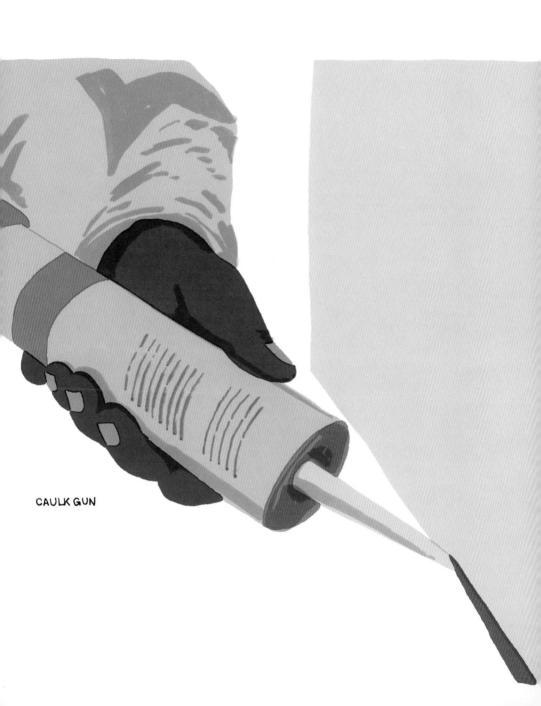

CAULK GUN

Clamps

It's often said that you can never own too many clamps. For most people who use tools and see the value in a good clamp, it's a statement that rings true.

A reason that clamp collections grow to fill whole walls is because clamps are so specific in their application. There is no "one size fits all." And because a project might call for one clamp, or thirty, you can easily end up justifying a vast array and number of clamps to have on hand, just in case.

C clamps (also known as **G clamps**) have been around for nearly a century. They work the same way as a bench vise, using a threaded screw to tighten a moving jaw against a fixed jaw. To withstand these forces, they need a sturdy cast-iron frame, which means they're heavy. Small and mid-range sizes are ideal for locking something to a benchtop and other similarly sized spans, but if you need to clamp anything wider than you can fan your fingers, you'll be getting into some pretty hefty clamps. In many applications, the C clamp is the perfect tool for the job, as few other clamps can match its force. (Large models used in steel construction can generate 40,000 pounds [178,000 N] of force.) The trade-off, besides their bulk, is that they require a lot of effort to set and remove.

Quick-grip clamps (or **trigger clamps** or **pistol clamps**, as they're also called) are comparatively new to the scene but now dominate much of clamp use. They're basically **one-handed bar clamps** (which

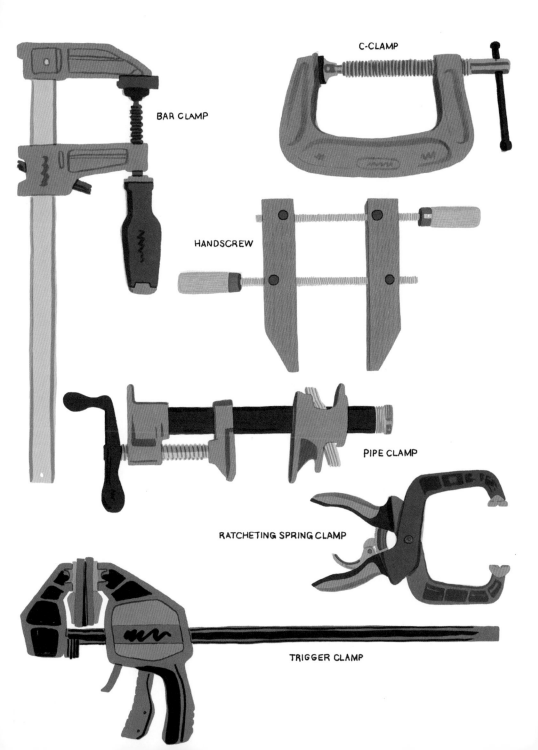

C-CLAMP

BAR CLAMP

HANDSCREW

PIPE CLAMP

RATCHETING SPRING CLAMP

TRIGGER CLAMP

is yet another name for them). They're lightweight and available in a variety of lengths, and their wide rubber pads are much better at not marring a surface than the button of steel found in a C clamp. In addition, you can also fully release their pressure via a quick-release trigger, which is a far cry from the effort required to loosen a C clamp.

The sacrifice with all this convenience is power. The vast majority of quick-grip clamps exert very little force, and it can be an incredibly frustrating experience having a workpiece constantly wiggling free from their grip. This is where it really pays to spend the money for a high-quality product. A clamp that barely clamps is worthless, so opt for the well-made larger versions. They'll clamp small spans just as well—and, given their negligible weight, won't be too much to lug around—and their more robust components will produce greater clamping force.

If you're in the market for significant clamping force across spans that are too big for a C clamp, look into **bar clamps** (or **F clamps**, as they're also called) as well as **pipe clamps**. Bar clamps are available in lengths that can reach from floor to ceiling. They adjust to size, usually with a pull of a tab and by sliding one face. Then they are tightened with a hand screw (the way you'd secure a C clamp). The better brands, such as Jorgensen and Bessey, can exert 1,500 pounds [6600 N] of force.

In woodworking, glued assemblies often require many bar clamps. It can turn into a pricey endeavor, and for this reason many woodworkers opt for a collection of pipe clamps instead. They have the same basic design as a bar clamp, but purpose-built bars are replaced with lengths of commonly available metal pipe. Pipes of any length can be more cheaply had, and this brings down the cost of building out a large clamp collection.

One style of clamp that sees little use in modern times is the **handscrew**. It's a traditional woodworking clamp made of two broad wood jaws joined by two adjustment screws. Handscrews have been mostly replaced by more capable technology, but there are still a few suitable applications. For starters, the handscrew's jaw angles are adjustable (all previously mentioned clamps have parallel jaws). Those extra-wide jaws can help in projects in ways that more narrow clamps cannot. And because the jaws are wood, they can be shaped for unconventional tasks, such as carving a V notch to better clamp dowels and pipes.

For smaller clamping needs, there are **ratcheting spring clamps**. These clips click down with ever-increasing tension and can be released with a trigger. They're the improved version of a regular **spring clamp**, which is, in essence, a large metal clothespin. Either one is good for clamping an area no wider than a finger or two and won't exert much force, but often that's all you need. There are also **1-2-3 blocks**, which are used by machinists in their fabrication setups. These palm-size precision ground blocks are threaded and can be clamped together (or to fixtures) in any number of configurations. They are surprisingly useful around the shop for odd tasks. **Locking pliers** with C-shaped forceps for jaws are another small type of clamp. They are typically used in metal fabrication and welding. They can exert more force than a quick clamp and be used just as easily, but they're quite limited by span and what their hooked jaws can reach around.

Come-Alongs and Chain Hoists

Come-alongs are hand-operated **winches** that use mechanical advantage to multiply force. They can be used for vehicle recovery, setting cable tension, or coaxing construction elements into place. They evolved from the first wooden winches used to snug up pontoon bridges in the Persian Wars some twenty-four hundred years ago (with the notable improvement of no longer using woven papyrus and flax as a winch line). If you've ever used a **ratchet strap** to lash down some cargo in the back of a truck, you've used a less robust version of a come-along.

Come-alongs are sometimes rated by the pounds or kilograms they can pull, but more often by the ton (which is convenient, as imperial and metric tons are roughly interchangeable). Most models range from a quarter ton to five tons in pulling capacity. A quarter ton can be plenty for getting the tension right while installing a large shade canopy over a patio, but to pull a truck out of a mud bog or to relocate a hefty section of fallen tree trunk, several tons may be required.

The most common versions of this tool use steel cable, which is spooled up into the frame as tension is taken up. The length of the cable is restricted by the spool size, so any long pulls must be made in stages. Two come-alongs can accomplish this, trading places as the cables of each are exhausted. Or progress can be captured with a rope or by **chock blocks**. But for longer continuous pulls, you need **rope ratchet pullers**. These come-alongs have knurled pulleys that pull along any length of rope. They're quite capable—so capable, in fact, that care must be taken to match the force of the come-along with an appropriately low-stretch and high-strength rope.

Most come-alongs are not rated for vertical use. It can be tempting to use these tools for the difficult work of, say, lifting a motor from a car's engine bay. They *seem* up to the task . . . but they're not. They don't have the ability to safely lower a load, so this is a job best left to a tool rated for it.

The few come-alongs appropriate for lifting and lowering will be marketed as **come-along hoists**, but a better tool for the job is the **chain hoist** (or, more accurately, the **ratchet lever chain hoist**). These operate on the same principles as come-alongs but can work vertically. They come in a similar battery of sizes and ratings. The chain does not spool. It feeds around the ratcheting pulley in the same manner that the rope-pulling come-along functions. The direction of movement is dictated via a lever on the handle (much in the same way that a ratcheting socket wrench operates), and the lever is cranked back and forth to move things along.

As with all modes of mechanical advantage that sacrifice time in the name of power, the going will be slow. But these are powerful tools with powerful possibilities; with hardly any effort, you can lift an entire truck clear off the ground. Just be sure to put it back when you're done.

Hammers

A **hammer** is simply a weighted head on the end of an arm. Given this low barrier of entry, many tools meet the qualification. You may have heard the old adage "Every tool is a hammer," the idea being that in situations that call for a bludgeoning, just about anything will do. In a pinch, sure, but most tools make a poor hammer and depending on the task, even the wrong *hammer* makes for a lousy hammer. If you match the right tool to the job, you'll see better performance, less arm fatigue, and fewer instances of accidentally bludgeoning that which shouldn't have been bludgeoned.

Claw Hammers

There are hundreds of types of hammer. Literally. Even back in 1867, Karl Marx noted that five hundred models of hammer were produced in a single British city, an observation that may possibly have informed some of Marx's thoughts on the bloat of capitalistic excess. But of all the hammer shapes, the one most readily conjured is that of the **claw hammer**. This is the hammer that comes in a basic apartment renter's tool kit for hanging a few photos. It's the same hammer that's slung on the tool belts of home builders (though they'd call those larger models **framing hammers**). The claw hammer is considered a general-purpose hammer, though one technically designed for pounding in nails and prying them back out.

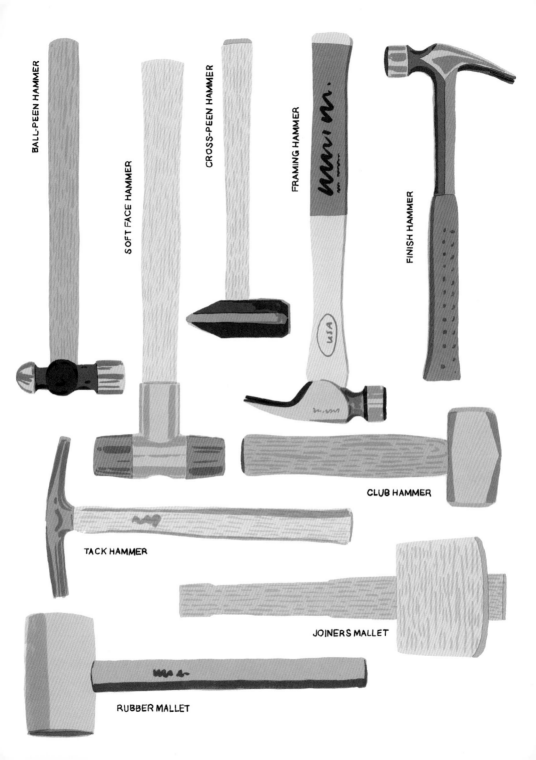

BALL-PEEN HAMMER

SOFT FACE HAMMER

CROSS-PEEN HAMMER

FRAMING HAMMER

FINISH HAMMER

CLUB HAMMER

TACK HAMMER

JOINERS MALLET

RUBBER MALLET

The claw on the back has a notch that's meant to slip around the shank of a nail to leverage it out. Often this is accomplished by the nailhead catching on the sides of the claw, but with a good hammer's sharp notch, the shank itself can be gripped—no nailhead necessary. Sometimes nails are buried flush in the wood and are difficult to remove. A well-made hammer tackles this by having a claw with sharply chiseled prongs that can dig into the wood and under the nailhead. Note that straight claws tend to outperform curved claws in this task.

Claw hammers come in a variety of weights, but 16 to 20 ounces [450 to 600 g] is plenty for general-purpose applications. In the hammer-intensive work of framing a home, a heavier hammer can paradoxically save on arm fatigue, assuming you learn to let the weight do the work and not your arm. But eventually all arms will tire, and for this reason quite a few pros swear by their **titanium framing hammers**. Manufacturer claims about energy transfer and vibration dampening are difficult to substantiate, but these hammers do have a few undeniable qualities that seem to support their ardent fandom and higher cost. Titanium has an improved

Practice makes perfect, but there are a few tricks that can help with removing a nail with a claw hammer. You can place a block of wood under the hammerhead to raise the fulcrum (the pivot point). You can also place the claw just ahead of the nailhead and strike the hammer's face with another hammer, driving the claw under the nail for better purchase. Some hammers have an extra nail puller built into the side of the head to provide an alternative vector of attack.

strength-to-weight ratio over steel, allowing for thinner handles that shift more mass toward the head, as well as longer handles than those of comparably weighted steel hammers. This means more speed at the end of the hammer and more energy transferred to the nail. And given the overall decrease in weight, they're less tiring to use. The only downside to these hammers—other than having to take out a loan to afford one—is that the mass of a traditional framing hammer is often called upon to knock lumber into place. Gentle nudges from a lightweight titanium hammer will be less effective. Its magic is really only apparent in the full-speed swings of nail driving.

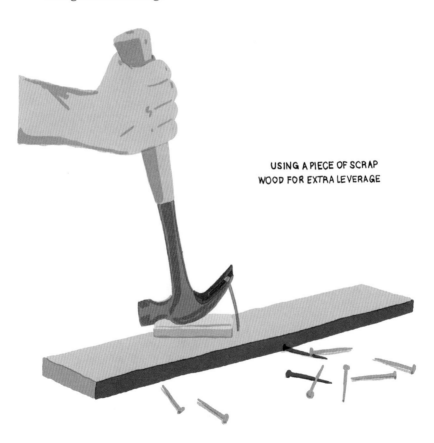

USING A PIECE OF SCRAP
WOOD FOR EXTRA LEVERAGE

The heads of claw hammers can vary. Framing hammers have a waffle texture for better grip on nailheads. If you see a notch in the top of the hammer face, that notch is called a nail setter. It's a magnetic slot that holds a nail and allows for the starting of the nail with one hand, as opposed to the usual two. A related tool that puts this mechanism to great use is the **tack hammer**, an upholstery hammer used for driving small tacks into furniture frames. Those little nails can be hard to grasp and start, and tack hammers have magnetic heads for this reason.

Given the number of manufacturers, the lengths, grip textures, head shapes, and weights, much of hammer selection comes down to personal fit. Go to the tool store. Try some on. Give them a swing. It's as personal a match as a pair of shoes.

Club Hammers and Mallets

Club hammers go by a few names: **drilling hammer**, **crack hammer**, **lump hammer**, and even **Thor hammer**. **Mini sledge** is a term often thrown around—as opposed to full-size **sledgehammers**, which are waist high with a 10-pound [4.5 kg] head. Mini sledges, by contrast, are one-handed tools weighing about 4 pounds [1.4 kg]. As hatchets are to axes, club hammers are to sledgehammers.

A mini sledge is the tool you'd use to strike masonry nails—or a masonry chisel, for that matter. In fact, if you just needed to smash some masonry, it's also the tool you'd grab. The stout steel head is unapologetic in its brutality. If you're in need of a more forgiving but still sturdy hammer, we have mallets.

Mallets are the lightweight, softer-touch version of club hammers. **Wood mallets**, sometimes called **joiner's mallets**, are the tool you'd use for coaxing wood joinery into place and then driving dowels through the connection. This is also the tool you'd use to strike the back of your wood chisel.

Wooden mallets are great for wood, but they lack the mass required for other applications. Also, they can be easily damaged by striking harder materials, so there are other mallets. **Rawhide mallets** are great for soft metal and jewelry. **Rubber mallets** can be used when a lot of mass is desired but you still need more give than a metal hammer would provide. Choose a model with light-colored rubber if you're setting hardwood flooring or doing other delicate work where you want to minimize any markings left behind—these are branded as "no-mar" or "nonscuffing."

Not to be confused with rubber mallets (or any of the other hammers with soft faces), a true **soft-face hammer** comes with two different plastic or rubber faces on the hammerhead and presents two tiers of cushion. They're less bulky than a rubber mallet, making them ideal for tight spaces. But another option for, let's say, knocking loose a bolt under a car hood would be a **plastic dead-blow hammer**. These mallets, designed with automotive and hydraulic work in mind, are filled with sand or metal beads to transfer force with little bounce. Then there are **brass hammers** and **bronze hammers**, which are mallets that can be used on harder metals without damaging the surface. These are often required in combustible work areas where steel-on-steel contact could spark an explosion.

There are a lot of specialized mallets. You could be excused for getting stuck on finding the exact right mallet for the job. But one doesn't need to know every process and every mallet—just the broad strokes to make a more appropriate match.

Peening Hammers

Peening is the process of working metal to strengthen it. Even though the process of peening is where these hammers get their name, they're used as general-purpose metalworking hammers. Of them, **ball-peen hammers** (or ball-*pein* hammers) are probably the most commonly encountered. They are used to fasten metal components by setting rivets and pins, rounding over the edges to fix them in place. They're also used to shape metal, strike cold chisels, and hammer **punches**. And they often meet the needs of someone who desires a lighter-weight mini sledge with a narrow face.

A **cross-peen hammer** trades a ball on one side of the hammerhead for a narrow wedge. Small versions are used in lightweight metalworking (such as jewelry), but they are also found in woodshops, as the blunt chisel end can be useful for driving tiny pin nails grasped between pinched fingers. Larger cross-peen hammers got their start in metalworking and are used to stretch and shape heated metal. **Straight-peen hammers** also owe their existence to metalworking. These hefty blacksmith tools have a wide face opposite a broad wedge. This is the classic blacksmith hammer shape, and if you've ever seen a statue or a union logo of someone about to beat a piece of iron over an anvil, it was likely a straight-peen hammer that he had hoisted overhead.

Heat Guns

A **heat gun** could easily be conflated with a hair dryer. They work the same and look about the same. A heat gun is an electric heating element in a pistol-shaped package that blows hot air out the front. But the difference between the two is that heat gun temperatures are probably more adjustable, and given that you can literally cook bacon with one, they are capable of much hotter temperatures. Most emit air streams in the 200°F to 1,000°F range [100°C to 550°C], though some can go a couple hundred degrees hotter. On some heat guns, the temperature settings are vague: low, medium, high. But on many models, you can select an exact temperature, which becomes critical in certain applications. Take removing lead paint, for example. You'd want to keep the temperature below 800°F [425°C], as above that the lead starts to vaporize and becomes a respiratory hazard.

Heat guns have found their way into a variety of applications for making and unmaking connections. They're a standard electronics tool, used on **heat-shrink tubing** to protect connections. They can be used to weld tarp seams and to weld and bend plastic pipes. Shrink-film packaging is cinched tight by way of a heat gun. And when it comes to rusty nuts and bolts, sometimes the easiest way to break them free is with heat; the expansion and contraction of the metal will loosen the threaded connection deeper than any solvent could penetrate. Heat guns are often called for in the stripping of old layers, such as when removing wallpaper or scraping off flooring and the mastics and glues that lie beneath (400°F to 600°F [200°C to 320°C] is about what you'd aim for here). Bumper stickers can also be removed from cars with the aid of a little heat, but care should be taken to use it sparingly and at low temperature lest you melt the bumper along with the adhesive. Likewise, with the right amount of heat and perhaps the assistance of a **dent puller suction cup**, heat guns can also pop damage out of a car's sheet metal. In other words, there are a lot of uses for these industrial hair dryers.

You have a few options if you don't have access to an electrical outlet. **Propane heat guns** are available for larger applications, and for smaller jobs, some **butane-powered soldering irons** have a heat gun attachment. Finally, cordless **battery-powered heat guns** are likely the most convenient option if you need a portable bacon fryer and don't object to spending a little more for it.

Impact Drivers

You might already own an **impact driver** and not even know it. Cordless drills are one of those power tools even the most casual tool owner might purchase, and manufacturers often offer discounted kits that contain both a cordless drill and an impact driver. So if you've ever wondered what the deal is with that smaller and noisier "drill" in your kit, here's all you need to know.

Impact drivers are designed for driving screws, drilling, and turning nuts and bolts. They look similar to a drill, but the most common models are generally more compact. These tools exert more force than a drill through the use of a spring-loaded hammer mechanism between the motor and the chuck. Here's how this works: As the chuck (and whatever bit it's holding) spins freely with little load, it does so with consistent speed and power (much like a drill), but as resistance is encountered, an internal hammer is automatically drawn back on a spring and released. As it releases, it drives into the back of the spinning chuck, simultaneously driving force into the fastener and increasing rotational force.

This happens dozens of times a second, producing a rapid clattering sound. The net result is that more force is applied, but in bursts rather than an even application. If using a drill is like pushing on a jammed door with your hands, using an impact driver is like kicking it open.

Impact drivers are also known as **impact wrenches**. In principle, impact drivers and wrenches are pretty similar, but there are some distinctions to the nomenclature.

Generally, the term *impact driver* is reserved for the smaller electric models, which these days are almost always cordless. They use a hollow chuck (also called a *collet*) that's spring-loaded and only accepts ¼ inch (6 mm) hexagonal drivers or drill bits that snap into place. Commonly sold kits of screwdriver bits will include both bits and magnetic bit holders that have a finger-pad-size depression near the rear of the hexagonal shank; that indentation means the bit is meant to lock into an electric impact driver.

The term *impact driver* can also refer to a less-common manual hand tool: a screwdriver meant to be hit by a hammer just as you'd strike a chisel. The tip rotates when struck, and when combined with a high-quality hardened driver tip and a bit of oil, this impact driver can be incredibly useful for removing a stubborn and stripped screw where the utmost control is required.

Impact wrenches (also called **impact guns**) generally have a square fitting on the end (called an *anvil*), like those found on socket wrenches and torque wrenches. The size of the fitting scales up with the power of the impact wrench, and anvils are available from ¼ to 1 inch wide. (They're sold in the same unit of measurement even in countries that use the metric system.) Impact wrenches are often found in manufacturing and auto shops. You may have seen (or more likely heard) them used to install the lug nuts on your wheels. In most auto shops, they're powered by compressed air, but corded and cordless electric models are also available, as their use extends beyond the lug nuts of a shop.

Impact drivers and wrenches solve some major issues with driving screws. First, drills lack the power to torque larger fasteners, and the effort to do so will quickly deplete the battery. Second, the constant rotational force can cause the driver bit to work out of the screw head (called *camming out*), and this will strip the screw. With impact drivers, you'll drive more screws per battery and find better purchase throughout the process. Plus, they have the added benefit of constantly hammering the tip of screws into the wood, meaning you'll penetrate easier.

That said, what you gain in power you lose in calm control. The action is not steady, it's erratic. Carefully boring a precise hole into wood is a job best reserved for a drill or a brace, not an impact driver. And while the triggers on impact drivers allow the speed to be modulated, it still takes a practiced hand to slow down and not over-drive a screw or bolt at the end of its run. On top of this, these tools are also very loud, and hearing protection is required. Most major manufacturers now sell a quieter impact driver or two, marketed as **hydraulic impact drivers** or **oil impulse impact drivers**, but they're a little harder to find. If you do seek one out, your ears and neighbors will thank you.

Nailers

Nail guns and **nailers**—they're the same thing. Even staplers are nailers, if they're powered in some manner. To qualify as a nailer, a tool should hold some sort of magazine of fasteners and power the driving, so you don't have to. Nailers are not a requirement for building. But if you're doing a construction project of any meaningful size, they can be a serious time-saver. And in some applications, such as hardwood flooring, nailers have become the de facto installation tool.

Unfortunately, nailers are highly specific. Most will shoot only a small range of fasteners. For a major project, you'd have to own three or four separate tools. For this reason, pneumatic kits are often packaged with multiple nailers. The assumption is that if you already have the **air compressor** to power them and you're in the market for a brad nailer, you probably want a finish nailer and a crown stapler as well.

Nailers can be categorized by their application: framing, finishing, flooring, etc. Before selecting a nailer, you first need to choose a nail (or other fastener). The correct nail for the job determines the nailer needed to fire it.

Framing Nailers

As the name would suggest, **framing nailers** are used for *framing*, or assembling the wooden studs and timbers that form the bones of many a house. They drive framing nails, but most are also capable of firing the smaller nails you might use in plywood sheathing, siding, or fencing. As with other styles of nailer, there are a handful of ways to power your framing nail gun. The most common type are pneumatic nailers powered by compressed air. This option tethers you to an air compressor, but the guns are cheaper than cordless options, and the power source is reliably consistent. If you plan to buy an air compressor, save yourself the frustration of hauling around a stiff, heavy hose and pay a little more for a thinner **polyurethane air hose**.

For those who wish to be entirely unencumbered, there are a few cordless options. First, we have **battery-powered framing nailers**. It takes a lot of energy to drive a framing nail at a rapid rate, so models can be hit or miss but are improving every year with better battery technology. The second option is **powder-actuated nail guns**. These use gunpowder to drive the nail, firing off what is essentially a firearm blank with each shot. These can be pricey to run, but they pack a punch. In fact, some powder-actuated drivers can fire fasteners into concrete or steel. Lastly, we have **gas-powered framing nailers**. No, they don't use gasoline like your car. These use canisters of compressed fuel to provide the power, along with a rechargeable battery to produce the ignition spark. The gas canisters will fire about a thousand nails and are often sold along with nails in boxed kits. They work well, but one issue to note is that prolonged sun exposure will cause the canister to overpressurize, leak, and become useless.

You'll need to match the nail gun to its ammo. Every nailer has different capabilities but also distinct requirements about the shape of the nail and how it's packaged. And it should be noted that you may find a nailer more reliably feeds one brand of nail over another, simply because of slight differences in manufacturing. Most will run a framing nail from about 2 to 3½ inches [50 to 90 mm] long; the larger nails are better suited for framing, and the smaller for plywood sheathing and fencing. For extra grab on the latter, look for ring-shank nails, which are ribbed and less likely to work their way loose. Strips of collated nails are laid out at a certain slant that's specific to each nail gun: 30°, 21°, etc. Buying the correct style of nail strips is like buying the right size coffee filter for your coffee maker. The diameters and the head shape of the nails will also vary. There are clipped heads (which have a crescent-shaped notch), full heads (which look like a normal nail), and offset heads (where the nailhead isn't centered on the nail shank). Besides matching the nails to the gun, this would also be a good time to consult building codes, which might specify the diameter, length, and number of nails and even which nailhead must be used for a particular task.

The depth that the nail is driven can be adjusted, and this will likely have to be done anytime you're switching between nails and projects. Set the depth so that the nail will be flush or perhaps just a bit below the surface. This adjustment might be as simple as turning a knob, or it might require a wrench and more involved finetuning at the tip of the nail gun. In the latter case, be sure the nail gun is without power.

Nail guns have safeties built into the tip that keep them from firing until they are pressed onto a surface. But like all mechanical things, they can fail. Mind where you point any nailer, but especially framing nailers. They deserve immense respect. The internet is littered with the x-rays of folks who didn't take their power seriously. You don't want similar carelessness to land you in the pantheon of medical curiosities like the railroad foreman Phineas Gage, whose two-hundred-year-long legacy is due to his having taken a **tamping iron** through the skull and barely living to tell the tale.

Palm Nailers

Palm nailers stand apart from all other nailers in that they don't require specialty fasteners. They have no magazine to hold collated strips of nails, and they don't *shoot* a nail into lumber. These tools are truly meant as an alternative to a hammer, pairing with the same common and box nails you'd be hammering in by hand, but perhaps in tight spots where a hammer can't be swung. Cordless battery-powered models are available, but more common (and often more compact) are those powered by compressed air. Most of the tool is held in the hand like a softball, with a piston-containing metal tube protruding from the front. This tube is slipped over a nail—or a nail is dropped into the tube and held in place via a magnet—and the nail is pressed into the wood. This pressure causes the piston to automatically cycle, impacting the nailhead dozens of times per second and sinking it to a set depth.

This ease of use, in a package hardly longer than many of the nails you'd use it on, makes it the perfect tool where access is limited—between studs, up under a cabinet, or around a corner to a spot that you can barely see, let alone reach with a hammer. These constraints are where the palm nailer shines.

As mentioned, other nailers use collated strips of fasteners specific to their proprietary feed system and sizing scheme. Palm nailers buck this by specifying the size of regular hammer-driven nails they are capable of driving. Be sure your nails fit your palm nailer before you get started.

How are nails generally sized? Good question. Much of the world uses metric, and these numbers are self-explanatory. You also might see lengths given in inches, or diameters measured in wire gauge. These too are still mostly straightforward, assuming you read up on how wire gauge sizes work in the "Snips" entry (page 128). But if you're in the United States, you might be facing a bulk bin of nails and confronted with nails sized according to the penny system . . . and like many outdated American systems of measurement, its perseverance is perplexing.

You can tell that nails are using the penny system because the size is punctuated with a letter *d*: 10d, 16d, and so on. Why *d*? The medieval English penny (where this system got its start) was designated by *d*—a derivation from the ancient Roman penny, the *denarius*. The number represents the size of the nail, but don't get excited—it's not in any way intuitive. It seems the penny number once referred to the cost for one hundred nails of a particular size. (So, one hundred 10d nails would cost ten pennies.) But it was impossible for those prices to remain static for long, so it soon became an abstract label even to the medieval English who invented it. Nonetheless, the naming scheme persisted, and now that number corresponds to a nail's length: 10d nails are 3 inches [7.6 cm] long, 16d nails are 3½ inches [8.9 cm] long, and so on. On a box of nails, the length is usually spelled out, but some palm nailers (and some building codes and builders) will call for nails by the penny system alone, so it's unfortunately necessary to be familiar with these vestigial labels.

Roofing Nailers and Siding Nailers

Roofing nailers and **siding nailers** are not all that different from framing nailers. They are available in the same variety of power sources, both hose-tethered and cordless. They fire a selection of nails—though these will be shorter and generally driven with less power, so as not to blow through shingles and siding. But the most notable difference visually is that the nails are stored in a large drum. These nailers are used in high-volume applications, and this coiled drum allows hundreds of nails to be locked and loaded. Owing to this easily noticed distinction, these tools are often referred to as **coil nailers**.

There's overlap in what these two types of nailers can be used on, but there are differences to be aware of in the nails they fire. Roofing nails tend to be smooth-shanked and shorter, and their nailer might not have a depth adjustment. Siding nailers are more like scaled-down framing nailers, firing longer ring-shank nails. A few models pull double duty and can drive both nail types, but in general you'll need to match the application and fastener to the correct tool.

Coil nailers are used for the highly repetitive tasks of installing sheathing, exterior drywall, vapor barriers, asphalt shingles, and siding, so speed is key. The winding-up delay experienced with some battery-powered framing nailers isn't as significant in cordless coil nailers, so you can expect to be able to shoot about three nails per second. That might be plenty for a lot of folks, but many pros still prefer the ability of pneumatic models to more quickly deliver a volley of nails with no hesitation.

It's important to maintain and lubricate all pneumatic tools, but it's worth emphasizing here, given the volume of nails these nailers are often tasked with throwing. They might be needed to sink several thousand nails in a workday, and jams or misfeeds not only will slow a job down but can also bring about enough frustration to cast a misbehaving roofing nailer off its roof. So, you'll want to lubricate the nail gun per the manufacturer's instructions. Generally, this means a couple of drops of pneumatic tool oil in the air inlet before use. You'll also want to regularly lubricate O-rings with compressor oil. This will keep them from drying out and cracking.

Crown Staplers and Flooring Nailers

While not technically "nailers," **crown staplers** fit neatly within this category. The mechanism and power sources of these tools are the same as their nailing counterparts and they're used in many of the same applications, often outshining traditional nailers.

Staples hold better than nails of comparable diameters. In part this is because they have two legs securing the material rather than one, but it's also because the *crown* of the staple—that bridge between the two legs—is wrapped over the surface that's been pinned down. Just how much distance this crown covers, along with the wire gauge of the legs, is determined by the capability of the stapler. They come in a few sizes. **Narrow-crown staplers** drive a staple that's ¼ inch [6 mm] wide and made of 18-gauge wire. These are sometimes called **finish staplers**. The staples are less noticeable and work well in locations that will remain visible, such as trim and latticework. **Medium-crown staplers** fire a 16-gauge staple with a ⁷⁄₁₆ inch [11 mm] wide crown. These would be used for attaching siding or shingles, putting down subfloor-ing, or assembling furniture. When it comes to upholstering the furniture, though, a more appropriate tool is an **upholstery stapler** that drives a thinner, narrower staple with lower force, so as not to break through the fabric. (There are also similarly sized staplers marketed as **carpet tackers**.) **Wide-crown staplers** use a 1 inch [25 mm] wide staple that is generally 16-gauge, and like all the previous staplers, they are available in a variety of lengths. By and large, you'd use these for similar applications as a medium-crown stapler but where even more holding power is desired, such as wall lathing, plywood sheathing, house wrap, roofing, and even truss building.

Flooring nailers are specialty staplers (though most are also capable of firing a fastener called a *cleat*) designed specifically for the process of installing tongue and groove flooring. These oddly shaped tools have an angled foot that serves two functions. First, it ensures that the angle of the stapler is perfect for driving a staple diagonally through the floorboard and into the subfloor. Second, the shape of the foot notches against the floorboard's tongue, which promises the perfect placement of the staple.

Installing flooring is tricky. It requires tapping floorboards into place to snug them up, and then attempting to secure them without losing any of that tightness. That can be difficult to do with a hammer and nail, so flooring nailers mitigate this by combining the fastening and the tightening into one action. Floorboards still have to be generally knocked into place with a **flooring mallet**, but when it comes time to really tighten up the boards and fire in the staple, the nailer is placed against the floorboard's tongue, and the nailer is struck with your mallet on a rubber bumper that protrudes from the rear. The hammer strike pushes the nailer against the floorboard, making a tight connection between the floorboard and the boards behind it, and at the same moment, the staple fires. Move several inches down the board. Rinse and repeat.

You might opt to load your flooring nailer with cleats, which are flat, L-shaped nails with serrated edges. There are a couple schools of thought when it comes to cleats versus staples, but it generally boils down to a belief that staples hold a little too well and don't allow for the movement of wood as it expands and contracts with humidity, while cleats do. The argument against cleats is that their larger size makes them likely to split thinner wood. This was more of a problem in the past; as cleats have found more favor, smaller sizes have been produced that reduce this risk.

Finish Nailers, Brad Nailers, and Pin Nailers

These are the little nailers. They're electric or pneumatic and fire a single, thin fastener. Don't let the nomenclature confuse you; the difference between these tools is only the diameter and length of the nail they drive.

Finish nailers use strips of 15- to 16-gauge [1.5 to 1.6 mm] finish nails. These are sturdy nails with a button head. Appropriately, they are used in finish work, crown molding, cabinetry, and some furniture building. Not unlike the improvement from framing nailers down to coil nailers, cordless battery-powered finish nailers are even more improved. In part, this is because there has been a demand for them and their portability, so they've been on the market longer than coil nailers, but mostly it comes down to the physics of driving a smaller nail and the demand for less power.

Brad nailers drive brads, which is just a name given to 18-gauge [1 mm] wire nails that have nearly no head. These are useful for trim work, especially on softer woods, and will leave an even less noticeable impression than a finish nail. The nails are just robust enough to hold on their own, but they are often used in combination with **wood glue**, holding the trim in place long enough for the glue to take over the task.

Pin nailers (or **micro pinners**) use 23-gauge [0.6 mm] wire pins. Like the fasteners used with finish nailers and brad nailers, these headless pins are available in a variety of lengths. They have virtually no permanent holding power, but they offer plenty of advantages. The small size punches a nearly invisible hole that doesn't require filling. For many projects where glue will be used, you can forgo the clamps and just pin everything together. Simply hurl pins into the workpiece, and then let the glue do the work. Because of the diminutive stature of the pins, you don't need to worry about visible holes or splitting wood. It's a difficult tool to make a mistake with . . . unless, of course, you take this casual sentiment too far and find yourself in need of a doctor to fish an errant pin from your hand.

Nail Pullers

There are plenty of ways to remove a nail. Most obvious is the claw on the back of your hammer, which will work fine for nails that are protruding. **Pry bars** and **crowbars** usually have slots for nail pulling, but they suffer the same limitations as a hammer and only work with easy-to-extract nails. So, as is often the case, the most efficient tool for the job is the one purpose-built for it.

First, there are **nail-pulling pliers**, which have wide, sharp jaws that can be worked in at the corners, but for digging deep you'll want something with more of a point. A **nail jack** is a similar plier-like tool that has a sharp point meant for pushing beneath the surface of the wood to get under a nailhead. **Finish nail pullers** are also a kind of pliers, but with generously rounded sides or a rounded horn that serves as a fulcrum for improved leverage. However, the length of all three tools and their inability to be beaten with a hammer limit the amount of force they can provide, so they're often relegated to extracting only smaller nails.

For quick work, better leverage, and the barbaric efficiency of something that can be pounded in with a hammer, there are **claw bars**. They go by a number of names depending on the location of the jobsite: **cat's claw**, **cat's paw**, **lamb's foot**, and **nail bar**. These L-shaped irons have a claw on both sides, allowing flexibility in the angle of attack and access. Both claws have sharp points, and the tool is meant to pair with a hammer. You start the claw a short distance from the nail, drive it under the nailhead with hammer strikes, and then pry the nail out. This will invariably splinter the wood that's pried up with the nailhead, but this can be minimized by driving the claw in a circle around the nailhead, severing the wood fibers. Some deluxe claws even have a circular punch to cut the grain around the nail, which will limit the blowout of the fibers.

If you really want to minimize damage, there's a more surgical vintage option. The **slide hammer puller** is a rarely seen relic of bygone days, though modern versions are still made by Crescent. It has two narrow pinchers that are placed on either side of a nail. The slide handle is slammed home a couple of times, driving the pinchers around the nailhead. Then when the puller is levered to one side, the pressure forces the jaws to close tight, and the nail is extracted. If you have to remove a bunch of midsize nails while salvaging a home's historical woodwork, this is the ideal tool.

A couple of other tools are useful additions to a nail-pulling arsenal: **Medical locking forceps** can be useful for working out thin nails, such as brads and small finish nails. **Rare earth magnets** are great for locating nails buried under years of caulk and paint. And a block of scrap wood is handy for placing under a hammer or prybar, either to get better leverage or to protect the wood that's being pried against.

Nail Sets

Nail sets are finger-length tapered metal rods used to countersink small nails just below the surface of the wood. This is done to keep from marring the wood with a hammer and to achieve a cleaner look. You might use a nail set on unfinished wood where you want nails to be less obvious, or on painted wood where you'll fill the hole with caulk and paint to completely deny the nail's existence.

Nail sets, also referred to as **nail punches** or **nail setters**, are usually purchased in a set with a handful of sizes. Most modern sets have colored rubber grips so you can easily differentiate sizes. Older models are simply chunks of steel with a knurled shaft for a grip. One end of the nail set is wide, to accept a hammer strike, and the other is narrow, to fit onto the head of a nail.

Nail sets are usually used to set a finish nail with its small pin head, but some can be used on thinner brad nails or even larger nails with flat heads. Some finish nails have a dimple that the nail set notches into, but on many nails the tip of the nail set cups over the top of the nail. In either case, the nail set being used should have a tip that is smaller than the nailhead so that it does as little damage to the wood face as possible.

As far as how they're used, the nail is driven most of the way in by hammer—a hit or two shy of flush so there's no risk of the hammer dinging the wood. Then the set is held against the nailhead, and the back of the set is struck with the hammer. A couple of taps will set the nailhead just below the surface for a polished look.

Nails that have been successfully countersunk are tough to remove. They can be dug out, but it will damage the wood. If you have access to the back of the wood, you might be tempted to remove the nails by hammering them from behind and back out the front, but this too will splinter the face of the wood. Instead, pull them all the way through with pliers. (Incidentally, this is the same way you'd remove a barbed fishhook from your finger, should you ever find your way to that misfortune.)

A related tool that's good to keep in mind is a **brad pusher**: a spring-loaded handheld driver for wire-thin brad nails. Most folks these days use a pneumatic or electric brad nailer, but for small jobs or an apartment hobbyist, these screwdriver-size devices will drive and set a brad nail with ease.

NAIL SET IN USE

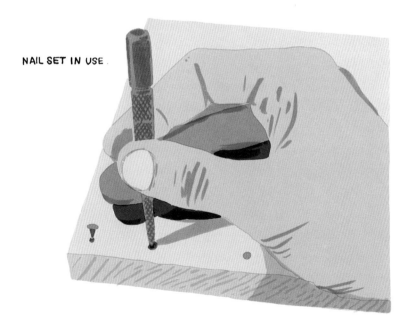

Pliers

Pliers are hand tools that use mechanical advantage to help you grip objects and apply leverage. If this sounds a lot like the description for a wrench, that's understandable, and you'll see some overlap between these tools. It's a line that gets blurry. But in general, wrenches don't require the user to grip two handles closed in order to grasp an object, while pliers demand exactly that.

It's a design descended from the two-handed **tongs** used in Bronze Age metalwork and the wooden cooking tongs that predated even those. There are many styles of pliers, but an easy way to divide the many variations is into three categories: fixed, adjustable, and locking.

Fixed Pliers

As you might guess, there's no adjusting **fixed pliers**. They can technically grip whatever fits between the jaws, but it should be noted that if the jaws are open too wide, most people will find they don't have the leverage to exert much force on the handles.

Narrow **needle-nose pliers**, one of the most common types of fixed pliers, are frequently used in electrical work for gripping and twisting small wires. Because of this, they tend to have wire cutters built into the jaws. They're also used to grasp small objects in tight places or to extract screws. (Similarly built **screw-removal pliers** have a specialized grip machined into the tip for this explicit

LINESMAN'S PLIERS

FENCING PLIERS

SLIP JOINT PLIERS

LOCKING PLIERS

END CUTTING PLIERS

NEEDLE-NOSE PLIERS

TONGUE-AND-GROOVE PLIERS

purpose.) Nearly as common are **lineman's pliers**. These stout pliers also owe their roots to the electrical work of cutting, pulling, and bending heavier gauge wire, but they've found their way into most toolboxes and kits.

There are more niche pliers: **Fencing pliers** are useful; they not only twist and cut wire but also hammer and pull staples. **Wire-twisting pliers** are the fastest way to twist wires together (other than putting the wire ends in an electric drill). **End cutting pliers** have two broad beaver teeth meant to grab and cut nails or to lever them out. These tools are used by both those who shoe horses and those who shoe humans. There are even **reversible snap ring pliers** that are used for spreading and removing retaining rings and clips.

Adjustable Pliers

Tongue-and-groove pliers, one of the more common types of adjustable pliers, are called by a few names, including **cobra pliers** and **water-pump pliers**. Most commonly, they're called **Channellocks** after a brand name so well known that the company that invented them eventually changed its name to align with its famous tool. The Champion Bolt and Clipper Company became Channellock in 1963, and they still produce a fine pair of pliers.

One thing to keep in mind with any adjustable pliers, such as tongue-and-groove pliers, is that with adjustments comes more play . . . more wiggle room. So, although the jaws of a single pair of pliers might accommodate pipes of varying diameters, they may not do it very well. To mitigate this, it pays to have a good pair. A well-made pair of tongue-and-groove pliers will grasp more securely than a cheaper model. A good pair, like those made in Germany by Knipex, will have sharper and harder teeth that dig into the round surface to resist slipping.

The other most frequently encountered adjustable pliers are **slip-joint pliers**. These bull-nosed models look a bit like lineman's pliers with their stocky build. They have a toothed recess paired with a flat knurled face, so that they might grasp both smaller and larger objects. There's a slip joint that allows them to be stepped open to accommodate larger sizes, though they don't adjust by much; the fronts of the jaws are flush when closed, and when the slip joint is adjusted, the jaws are perhaps a pinkie's width apart. But it's enough to accommodate most odds and ends found around the house and saves on carrying multiple pairs of fixed pliers. For this reason, slip-joint pliers are usually included in basic home and vehicle tool kits.

Locking Pliers

Not unlike tongue-and-groove pliers, **locking pliers** are referred to by the names of the companies that invented them. In the United States, they're called **Vise-Grips**. In the United Kingdom, they're known as **Mole Grips**. Both designs use the same principles. When you squeeze the handles of the pliers, just as maximum pressure is exerted, the hinge pivots over a fulcrum and drops down the other side, locking the handle in place. A threaded knob at the end of the handle can be adjusted to position the fulcrum and change the jaw size, but it can be a little tricky to use until you're used to it.

The method of fiddling with the size might look like this: Close the pliers all the way and compare the opening of the jaws to the object you want to grip. Let's say that object is a pipe. Rotate the threaded knob to adjust the jaw opening until it's about the size of the pipe. Then open the pliers and clamp down on the pipe, squeezing past the point of tension until the handles lock in place. Was it easy? Too easy? The more difficult the pliers are to lock in place, the more securely they're locked. If it was too easy, press the release tab on the handle to spring the pliers open, tighten the knob just a little, and try again until the handles lock into place with authority.

Locking pliers are simultaneously indispensable and one of the worst tools to give the uninitiated. They can apply immense force and stay put. For this reason, they are uniquely capable of solving problems, including grabbing nuts and screws that have been stripped completely beyond use with any normal wrench or pliers. But on the flip side, they're misused by lazy users who don't care to find an appropriately sized wrench, which is often the very reason why that nut is mangled in the first place.

The biggest difference between models of locking pliers is the size and shape of the jaws. All their mechanisms are basically the same, but the utility of a strong one-handed clamp has led to all sorts of adaptations. There are needle-nose jaws and wide pipe jaws, as well as big flat paddles used as a **handheld sheet metal seamer**—a device that makes bends in sheet metal. There are long hooked forceps made to fit around large objects and grip at the tip, and these are used as welding clamps to keep things in place during that process or for clamping a workpiece down on a table.

Rivet Guns

In industrial applications, rivets have fallen out of favor and been replaced by stronger mechanical fasteners, so these days **rivet guns** are mostly relegated to lightweight projects, where they're used to fasten blind rivets, also known as *pull rivets* and *pop rivets*. (Pop was the brand name of the original manufacturer, now owned by Stanley Black & Decker.)

Rivets have been engineering staples since the Victorian era. Metalwork from that period, from boilers to bridges, is still in use today, dotted with the raised bumps of dumbbell-shaped fasteners. Ironworkers tossing hot ingots to one another and pounding the malleable steel with a **pneumatic hammer** is an iconic image of early-twentieth-century industry. But those rivets have fallen out of favor. They were dangerous to handle and hard to install, and because they were heated during installation, they could not be heat-treated. (Heat treatment, the process of heating and precisely cooling metal for improved strength, would be undone with reheating and imprecise cooling.) In places where mechanical fastening is still required, bolts have largely replaced rivets.

But not everywhere. Sheet metal fabrication still uses plenty of rivets where lightweight fasteners are desired to pair with the equally airy metal. The skin of aircraft is generally riveted together. Metal signs are riveted, and plastics too on occasion. The most common place around the house to find rivets is in the connections between metal rain gutters and on the sheet metal of some appliances.

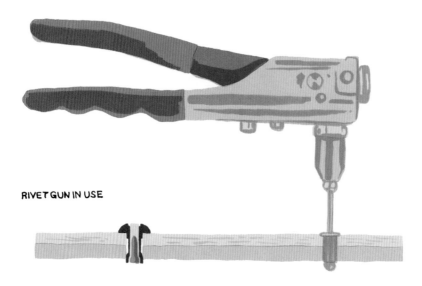

RIVET GUN IN USE

These lance-shaped bits of aluminum or steel are meant to be inserted through only one side of a drilled hole, fat end first (hence the term *blind*), with their thin stem (called a *mandrel*) housed inside the rivet gun. The handle is squeezed a few times, drawing the stem deeper into the rivet gun as a bulb mushrooms on the backside of the rivet. The mandrel eventually pulls through (or breaks off, depending on the rivet's design), leaving the rivet firmly compressed against both faces of the material being fastened.

There are a few considerations for this otherwise simple process. For starters, you need to match your rivet material to your metal. Mixing metals produces a reaction known as *galvanic corrosion*, which will cause the connections to rot away. It's also important to drill a hole exactly the same size as the rivet and to select a rivet whose length, when compressed, is adequate for the material stack that's being fastened. A rivet that's too long will never fully compress. And while on the subject of rivet size, there are various thicknesses of mandrel. To accommodate this, rivet guns have a few fittings with mandrel holes of different gauges. Generally, the extra fittings and the wrench used to install them are stored on the handle. A neat option to be aware of are internally threaded rivet inserts meant to accept bolts, called *rivet nuts*; you'll need a special **rivet nut tool** or adapter to install these.

Manual rivet guns are elegant machines. They're simple, hardly more complex than a pair of pliers. And they're cheap. For the average user and sporadic projects, nothing more is needed. But if you've recently purchased a kit aircraft or are looking to get into the gutter business and value your hand strength, there are a few other options. Dedicated **battery-powered rivet guns** are the most obvious solution, along with the air-powered **pneumatic rivet guns** found in most sheet metal shops. If you're looking to experiment with something cheaper, though, you might look into **rivet gun drill adapters**, but be warned that these work better on light aluminum rivets and can generally be temperamental, so your results may vary.

Screwdrivers and Drives

Screwdrivers are hand tools used for tightening and loosening fasteners. The most basic screwdrivers are merely handles with driver ends that match up to conventional screws, like Phillips-head screws and slotted screws. But a screwdriver handle could also be fitted with a hexagonal Allen wrench tip, a proprietary security bit, or even a socket (these are usually called **nut drivers**). Basically, regardless of what's on the business end, even if it's meant to turn a bolt or a valve, the tool might still be called a screwdriver.

Many of the screwdrivers used today are just handles that hold replaceable tool tips, called **bits**. These come in a variety of packages. There are **multibit screwdrivers** with bits for various screw drive styles stored in the handle. There are **right-angled screwdrivers** and **ratcheting screwdrivers** that use these bits. There are also time-saving **electric screwdrivers**, which are essentially underpowered drills. On that note, you can also use a screwdriver bit in the electric drill or impact driver that you already have, and in fact this is one of the most common approaches. **Screwdriver bit kits**, sold alongside drill bit indexes, have an assortment of adapters, magnetic extensions, and replaceable bits that fit into your drill, impact driver, or handheld screwdriver.

So if a screwdriver is just a handle (or motor) to turn a collection of bits that tighten or loosen various screws, there's not much to say about the screwdrivers themselves. Where things get interesting is in what the bits mate to: the drive styles of those screws, also called the *screw's drive* or the *drive type*. The drive style is ultimately what affects the performance of a screwdriver and is where all the nuance lies. What follows is a breakdown of the most common screw drives found at the hardware store and thus the most common bits in multibit screwdrivers and in screwdriver bit kits.

Slotted

Also called a **flathead** and probably the most familiar screw drive type, the **slotted** screw drive was the first to be invented a few hundred years ago. You'll find screws with a slotted head in old construction or on antiquated machinery, as well as on cabinet and door hardware.

Slotted screws are generally low-grade, multipurpose screws, though not always; slotted screws on sewing machines, firearms, fishing reels, and other well-cared-for machinery are often well made and demand a better fit than you'd get from a general-purpose slotted screwdriver. Most slotted screwdrivers and slotted screwdriver bits are tapered to accommodate different sizes of slotted screw, but these one-size-fits-all tools can *cam out*—that is, they might twist free as torque is applied—and damage the slot of the screw. To combat this, **hollow ground flathead screwdrivers** (often sold as **gunsmith screwdrivers**) provide a flat blade with parallel sides that notch tightly into the full slot of the screw. This increase in contact area provides more torque, which can be

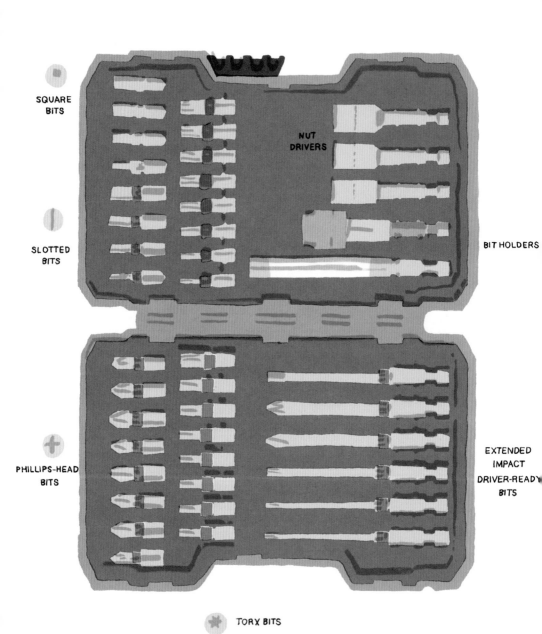

SQUARE BITS

SLOTTED BITS

NUT DRIVERS

BIT HOLDERS

PHILLIPS-HEAD BITS

EXTENDED IMPACT DRIVER-READY BITS

TORX BITS

crucial if a screw has been secured with **thread-locking fluid**, such as Loctite. That positive contact will help prevent camming and maintain the integrity of the screw.

Screwdriver bit kits usually have slotted bits in a couple of different sizes to accommodate large and small screws. But you might also have a stand-alone screwdriver for this drive style. Your classic slotted screwdriver seems to be one of those omnipresent general-purpose tools that's passed down through the generations. And you can be excused for using one to pry open the occasional can of wood stain. It's at least a step above using a knife or your car keys.

Robertson

The Robertson screw drive is almost exclusively referred to as the **square drive**. It was invented by Peter L. Robertson, of the Robertson Screw Company. The lore is that Robertson, a Canadian inventor, cut himself with the blade of a flathead screwdriver that slipped off a screw during a demonstration of a new spring-loaded driver he'd created. It was a eureka moment that led him to invent the more secure screw drive he felt the world needed. In 1908, he designed a fastener that featured a square socket. This design was vastly superior to the slotted screws that came before and it should have become the global standard, but it didn't for reasons that we'll soon get to.

Robertson's screw drive style remains largely unchanged and is still quite common in Canada. In other countries, the Robertson drive style is less frequently used, but it can be found on many stainless screws and on an assortment of sheet metal and wood screws. For this reason, screwdriver bit kits will usually include the most common sizes of square drive bits: #1 to 3.

Cruciform

There are a handful of **cruciform screw drives**, but the most common—and in fact the most common screw drive type in the world, period—is the Phillips. You might hear a **Phillips-head** screw also called a **cross-head** screw. A few styles are so visually similar that they're easily conflated with a Phillips, but a true Phillips-head screw, and the screwdriver bits that drive it, features a single cross-shaped recess that tapers inward at the bottom.

> Twenty-eight years after Robertson invented his square drive, he was still finding it difficult to convince mass manufacturing to pivot to it. In large part this was due to his stubborn nature and an inability to strike licensing deals. Henry F. Phillips did not share this inability. His cruciform screw drive design didn't provide as much torque as Robertson's, but he was a more adept salesman. He brought the design to American Screw Company, which in turn put it into the hands of automobile manufacturers. It caught on like wildfire within the industry, not because of its superior performance to the square drive but in fact because of its *poorer* performance. Auto manufacturers actually preferred that the Phillips drive would eventually cam out. They viewed it as a feature, not a bug, that prevented the overtorquing of screws on a quickly moving assembly line.
>
> Automobile manufacturing was a dominant force that developed the very concept of assembly lines and shaped many industries, so the adoption of the Phillips screw drive rippled out. Today, it's the most common screw drive in the world, and yet, because it's easily stripped, it's also one of the worst.

Applying firm pressure when using a screwdriver on this style of screw drive will help avoid the stripping that is somewhat inevitable, as will using the correct size of screwdriver or screwdriver bit. They're numbered 0 to 5, with #2 being the most common. There are also Phillips variants, so be sure to match the style of the driver to the screw. A faintly stamped X over the cross indicates a **Posidriv** screw, and a slash across the cross with a small square recess in the center is a **Supadriv**. All these Phillips drive evolutions provide marginally more purchase than the original, but only if you use the correct bit.

Ultimately, the Phillips drive is a flawed design and one that society seems to be stuck with. There are better ones available. The aforementioned four-sided square drive certainly works better. And there are a hundred niche fasteners with improved performance characteristics, but the most common better-performing drive styles are five- and six-sided.

Hexalobe

The six-pointed hexalobe screw drive is most commonly referred to as a **Torx** or **star drive**. It's been around since 1967, but it has required several decades to really catch on. The design borrows from the hex socket (a common machine-screw drive style), recognizing that more facets allow for increased torque and less wear. The star-shaped socket builds on this with six rounded lobes that create even more surface area and a shape that can't easily be stripped out.

The positive connection and lack of cam-out means less wear on both screw and screwdriver bits. And because the driver bit doesn't have to be carefully lined up with only a single notch or two, it easily slots into place in almost any position. Thus, this drive style has become a time- and money-saver that's seen wide adoption.

Many machine screws, structural and trim screws, and almost all deck screws feature the hexalobe drive. The most common sizes are included in bit kits and are labeled T15, T20, T25, and so on. It's also become an increasingly popular security screw drive, with specialized bits that notch over an additional pin centered within the star. Note that hexalobe screws shouldn't be confused with the less-common five-pointed **pentalobe** screws you might find on your laptop or smartphone.

The increased performance and decreased wear of high-quality hexalobe screws like those manufactured by GRK are a world apart from other screws, and their reusability is well worth the difference in price. This becomes even more apparent in home remodeling, where adding and removing framing is made ever smoother with T25 framing screws. These are easier to install (and uninstall) in tight quarters than nails and are worth their weight in gold when you realize that a stud you just replaced needs to be relocated a little to the left.

Soldering Irons

Soldering irons are self-heating tools used to melt metal for the purpose of fusing two objects together, most often wires, circuit boards, or components. They can also be used to bond stained glass and jewelry.

Basic soldering irons don't allow for temperature adjustment and are little more than a heating element at the end of a cord. Improved models include a base station with temperature control, a stand, and a sponge tray for cleaning the tip of the iron. If access to power is an issue, there are even **butane-fueled soldering irons**. All of these have replaceable tips, in part because the tips will corrode eventually, but also to accommodate different needs: fine points for small connections, broad chisel points for larger ones.

To solder properly, you need a few supplies beyond the iron. You'll need **solder wire**. **Flux core** tends to be the easiest to work with. A damp sponge and steel wool or some fine sandpaper are also necessities. Soldering is delicate work, and to clamp the work-pieces together, a **third hand** (sometimes called a **helping hand**) is useful; this is a block with movable arms and clips that can be used to hold pieces in place as you solder them. **Flux paste** would also be good to have, even if your solder is embedded with flux.

What's flux? Flux is a paste that makes soldering possible; without it, you'll never solder a decent connection. This is because metals oxidize and solder won't bond to oxidized metals. In fact, metals oxidize even more rapidly when heated, so you can see the dilemma. Flux solves this problem by removing and preventing oxidation; it also allows the solder to flow more easily.

To begin soldering, turn the iron on and adjust the temperature if possible. The proper temperature for a soldering iron depends on the tip and the materials, but a good temperature range is 600°F to 700°F [320°C to 370°C]. Clean the tip of the soldering iron with a damp sponge, and then touch a bit of solder wire to lightly coat it. This is called *tinning the tip*. It will help conduct the heat and prevent oxidation on the soldering iron.

Before applying solder, lightly sand the workpieces. Brush them with flux. Then hold the iron like a pencil, with the tip against the workpieces where you intend to make the connection. Touch the solder wire between the iron tip and the components to melt a bit of solder and better facilitate the heat transfer between them. Then apply solder to the far side of the component, opposite from where the iron is touching. The solder will be attracted to the heat and should wick around the whole connection, making a properly soldered joint. If solder doesn't flow around the entire connection, the parts aren't hot enough or should have been cleaned more thoroughly.

Once a proper connection is made and you're satisfied with your work, you may need to clean the joint with alcohol and a brush to remove flux residue. Some flux or solder will specify no cleaning, or clean only with water. Check the package.

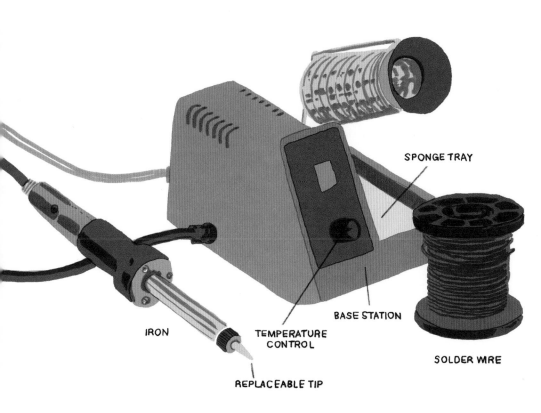

SPONGE TRAY

BASE STATION

IRON

TEMPERATURE
CONTROL

SOLDER WIRE

REPLACEABLE TIP

SOLDERING IRON STATION

Staple Guns

Staple gun staples are U-shaped metal fasteners, not unlike those found in the stapler on your office desk. Those used in the shop (or on a construction site) are more robust, with either a rounded or flat top and sharp points on their legs. That familiar staple shape makes them useful for securing skinny materials, like a length of cord or some holiday lights. They're also ideal fasteners for tacking down thin layers, such as water-resistant house wrap, where nail-heads would pull through.

Many people are familiar with the classic **manual staple gun**. These blocky, steel, lever-action models have been around since 1929, when they were first introduced by the Arrow company. They haven't changed much since then, though some manufacturers sell improved **forward-action staple guns**, which move the lever to the front of the tool, meaning that force is applied directly over the staple rather than behind it. Regardless of who manufactures your modern staple gun, or how its lever is oriented, the size of the staple it uses is still designated by a proprietary naming system developed by Arrow. Most staple guns fire a utilitarian T50 staple, but within that size a few different lengths can be purchased. T55 guns, an offshoot of T50, accommodate even longer staples. T25 guns fire an arched staple meant for low-voltage wire. JT21 staples are for light-duty applications. And on it goes.

What does T50 actually stand for? Good question. In a 1995 trademark case, Stanley Works (a competing staple gun manufacturer) argued that the T was in reference to the gun being a "tacker" and the 50 meant that it fired staples with a wire thickness of about 0.050 inches. Thus, T50 was a real-world size designation free for anyone to use and no licensing fees need be paid to Arrow. Arrow, arguing that the name was theirs to use alone, insisted that they'd chosen the T50 name completely arbitrarily. You might be inclined to side with Arrow—after all, the coded names for all the other staples are an inconsistent mess. But in this instance, the courts saw it Stanley's way.

For staple jobs that require less exact placement and a lot of staples, **hammer tackers** are the tool for the job. These stick-shaped staple guns are loaded with the same strip of staples found in other manual guns. They're whacked against a surface, and the impact drives a staple on each hit. Tacking down roofing paper or the plastic sheeting for insulation is an ideal use of these tools.

Because staples have improved holding power over nails of similar size, they have found their way into a lot of applications and into bigger sizes and more capable guns. **Electric staplers** and **pneumatic staple guns**, sold alongside other nailers, are used in woodworking, installing carpeting and hardwood flooring, upholstery work, and construction (see page 168 for more details).

Taps and Dies

Taps and **dies** are the tools used to create (or clean up) the threads of bolts, nuts, and other threaded screw holes. They're generally assembled in kits, as there are a wide assortment of diameters and thread counts to accommodate. Many of these kits contain lower-quality parts than individually purchased components, but they're perfectly adequate for the novice. And while threading your own nuts and bolts might seem like a pastime reserved for only the most dedicated of fabricators, these tools actually have surprising utility and ease of use.

Let's first separate taps and dies. A tap looks like something between a screw and a drill bit: a fluted, threaded rod with a tapered point and a square shank on the rear. Taps are used for cutting the threads in nuts and threaded holes—what's technically known as the *female* side of threaded mechanical connections. Dies are wheel-shaped and have a center hole that's threaded. Notches in this threading provide the cutting edges required to carve threads onto would-be bolts—the *male* end of threaded mechanical connections.

Because a vast assortment of premade bolts can be readily purchased at the hardware store, tapping holes tends to be the more common of the two processes. For example, a previously tapped hole might be ruined, so you drill it larger and tap fresh threads into it. Or a bolt might be added to an existing object—say, retrofitting a car with a new roof rack add-on—and new holes must be tapped to receive a screw. Also, hobbyist makers and fabricators often need to tap threaded holes to assemble their creations with store-bought screws.

There are three main styles of tap, all of which are available in every desired diameter and thread count. The first is the **tapered tap**, which has a generous taper to its diameter. Tapered taps are good for thin material, as they're easy to start and turn, but they will have to be run all the way through to attain full-diameter threads.

Then comes the **plug tap,** as it's known in North America. (In Australia and Britain, it's known as the **second tap**.) Plug taps are the most common style. They have a gentle taper on the tip—enough to easily start the tap into a hole—but quickly attain full diameter. For most folks, these are the only taps that are required.

The last style is the **bottom tap**, which is used for tapping blind holes—those that don't extend clear through a workpiece to the other side. Because they have virtually no taper, bottom taps can cut threads to the bottom of a hole, but they can't actually start a hole. Blind holes need to be threaded initially with a plug tap so that a bottom tap can then be inserted to complete the threading.

To tap a hole, you first need a hole. If you're unsure which tap is required to match a screw, use a **screw-pitch gauge** to identify the thread profile. Then match the appropriate drill bit to the tap. These charts are online and usually also found inside the case of the tap and die kit itself. Select the correct drill bit, and drill the hole. It's not always critical to drill a straight hole—like when boring through a few layers of automotive sheet metal—but with thicker stock, you'll want to keep that hole straight and use a drill press if possible.

Next, attach the tap to the **tap wrench**. Some kits include larger wrenches, which should be avoided with small-diameter taps. You need as much tactile feedback as possible, and feedback is diminished in the sloppy connection between an oversize wrench and a small tap. Appropriately sized and well-made tap wrenches, like those made by Starrett, are worth procuring for their simplicity and tight tolerances.

Apply some **cutting oil** to the plug tap. You can't use too much. Start to twist clockwise, and soon you'll feel the threads starting to cut. Keep the tap straight (a **tap guide** will help with this), and twist one rotation clockwise, then rotate half a turn counterclockwise to break the chips and clear the *swarf*—that is, the shavings that are piling up. Repeat the pattern of one rotation forward, half a rotation back. (This is a rule of thumb, and it can vary based on how much resistance is felt.)

Run the tap its full length, then carefully unthread it and clean the hole with compressed air before running the tap through once more to clean the threads. That cleaning process can also be done on pre-existing threads that are a bit dirty or beat up. Tidying up old threads is called *chasing*, and there are specialty **chaser taps** that are made with softer steel so you don't accidentally cut new threads, but regular taps will work fine if you use them carefully.

Now, on to **dies**. Cutting threads with a die is called *threading*. The process of threading a rod, or cleaning up the threads on an existing bolt, is very similar to the process of tapping.

To prepare a rod for threading, you'll want to grind a slight bevel onto it, just as there's a bevel on taps and on many bolts. Check the rod size with a pair of calipers, and select the appropriate die.

Existing bolt threads can be measured and checked against a screw pitch gauge. Then secure the die within a **die wrench**.

Mount the rod in a vise, and apply a liberal amount of cutting oil to both the rod and the die. Now you're ready to start cutting. The process is the same as with a tap: one rotation clockwise, half a rotation back. When you've finished, clean the threads with compressed air, and run the die back through a couple of times. **Chasing dies** are available for the explicit purpose of chasing male threads, such as when cleaning up spark plug threads to remove carbon buildup.

Here's the important takeaway from this discussion of taps and dies: Use more cutting oil than you think you need, go slow, and don't force it. Remember this advice, because snapping off the ultra hard metal of a tap where you'll never get it out can be a very costly endeavor, both financially and emotionally. In fact, your best bet, if you're new to tapping, is to buy a cheap kit and some disposable material to practice tapping with. Feel free to do so with vigor. It might break and that's just fine. Cutting threads is a process done by feel, and it's well worth the time and money to get a sense of the appropriate force by sacrificing a tap and some material.

Welders

Welding is a fabrication process in which parts are fused together via heat, pressure, or both. Welding is most often performed to bind metal components, but plastics can also be welded, as can wood, though the latter is more of a science experiment than a practical process. Metal welding is the focus of this section.

Welding Safety

Welding is fraught with dangers: hot sparks, arcs that can blind you, and explosive gases. Many community colleges and maker spaces offer classes that will give you a sampling of the welding processes, and you'd do well to learn the finer points and safety protocol from a professional.

That said, some processes like MIG welding are much easier to tackle than others. If you go this route, you'll need a welding jacket, gloves, and a helmet with built-in mask. The tint of the mask's eye protection is critical, and even if you're not the one welding, don't ever look at a welding flame or arc without eye protection if you enjoy your vision.

You'll also want a clean work area and a stable position to weld from. Everything needs to be in place and steady because you'll be behind a very dark mask and focused on your weld bead. As such, you'll probably want to clamp what's being welded in place. **Magnetic welding squares** and other jigs can temporarily hold an assembly until you tack it in place with some welds.

There are nearly as many welding techniques as there are metals to weld. Metal type and thickness, the environment, and the desired appearance of the weld all factor into which machine, materials, or techniques will be used, but all techniques share three elements: heat, filler material, and shielding gas or flux.

Oxyacetylene Torches

Most welding techniques require an electrical arc, but an **oxyacetylene torch** (the oxy stands for "oxygen") is just that: a torch. Torches also have uses beyond welding, so their utility and the portability of two tanks on a cart make them prized tools in remote sites without access to electricity.

The most common nonwelding use of a torch is for metal cutting. Torches can also be used for brazing, soldering, or heating metal to bend or hammer it into shape.

Let's look at the process of using a torch to weld steel. You'll want the correct tip on the torch, and full tanks of both oxygen and acetylene. Set the tank pressure regulators to 5 psi [50 kPa]. Open the acetylene valve on the torch about a quarter turn, and ignite the flame with a **flint torch igniter**. Now adjust the acetylene valve: too little and the yellow flame will be sooty, too much and there will be a gap between the torch tip and the flame.

Once that's set, add in oxygen. As you do, you'll see two blue cones of flame form within the larger yellow one you started with. Continue opening the oxygen valve until the larger blue cone disappears into the smaller cone, leaving a lone blue spike of flame surrounded by a larger yellow plume. If you add too much oxygen, that blue cone will start to shrink, and the hissing noise will get louder.

To start welding, hold the torch at an angle away from the direction of the intended path, just above the workpiece. Use your dominant hand for the torch, and hold the filler material (called a *gas welding rod*) in the other hand. Use the torch to heat the base metal until a molten puddle forms. In torch welding (and really all welding), it's important to remember that you are melting the base metal and adding filler material to the puddle; it's the puddle that melts the filler, not the welding torch. Start pushing the torch forward, moving the puddle along the path, feeding filler material into the puddle as you go, leaving a bead cooling in your wake. Once you're done, extinguish the torch by shutting off the acetylene first, then the oxygen. Don't be startled by the sputtering pop.

Stick Welders

The technical name for the next process is *shielded metal arc welding*, but most people just call it *stick welding*. Here's where the technical name comes from: The sticks, consumable electrodes that look like large sticks of incense, are composed of a *metal* filler material that's surrounded by a coating of flux that *shields* the welding process. The welding is possible because the stick is attached to the positive wire of a machine that generates electrical current. As the stick is brought near the workpiece (which is attached to the negative wire), an electrical *arc* is sparked, causing immense heat that melts the workpiece metal and filler rod.

With **stick welders**, the correct electrical power and stick must be selected for the metal and thickness being welded. A good ground clamp connection is critical, and you must follow general welding safety practices.

The stick welding process goes something like this: First, secure the welding rod in the positive wire's clamp. You'll often hear this clamp referred to as a *stinger*. Hold the rod vertically on the intended weld path, and lean it 10° in the direction you'll be moving, so that you'll be dragging the tip as you go. Then strike an arc, either by tapping or by swiping the tip as you would strike a match. The idea is to warm up the tip and eventually strike an arc on the spot you're welding where the tip can then be kept close enough to maintain that arc.

Once you've struck an arc, drag the tip while simultaneously feeding the consumable welding rod into the puddle. For those new to welding, the rod has a tendency to stick to the workpiece. This usually results from failing to maintain the correct distance or using too low an amperage setting on the welder. If the rod sticks, unclamp the stinger and break the rod free by hand. Then reattach the rod to the stinger and continue on with the weld.

Stick welding has a few advantages over other types. For starters, it's ideal for outdoor work where wind can be problematic for gas-shielded welders. Stick welding also tolerates dirty metals, as well as those that are quite thick, thus it's perfect for the remote work of repairing an old farm gate or a tractor. This is why many stick welders are gasoline-powered.

Here's a neat trick if you find yourself marooned without a gasoline-powered welder but with a few old cars and some welding sticks: You can actually stick-weld with two or three car batteries wired in series.

Stick welding has a few disadvantages. It discharges a lot of vapor, as well as spatter. Slag will form on the outside of the weld as excess flux cools and hardens, and this will have to be knocked off with a **chipping hammer**. Even then, the welds won't be as clean as those produced by other processes.

MIG Welders

MIG (rhymes with *fig*) stands for "metal inert gas." It's become the commonly used name for all gas metal arc welding, regardless of whether the gas being used is actually inert.

MIG welders use electricity in combination with pressurized gas and an automatic feed of filler material. The gas is doing the job of the flux in stick welding, shielding the welding process from the atmosphere. The welding machine provides an electrical arc for the heat that's required, and it also feeds filler material directly out the tip of the welder gun.

Besides its relative ease of use, MIG welding has advantages over some other types of welding in that it produces cleaner welds with no slag. It can be used to weld steel, stainless steel, and aluminum in a wide variety of thicknesses. Also, many MIG welders can use flux-core wire, which allows you to forgo gas altogether. The exact capabilities of a MIG welder will depend on the filler wire, power setting, gas type, and feed rate. Tables that cover all of this are standard issue with the purchase of a welder.

To use a MIG welder, you must first prep the surface material. That means cleaning it with a **welding brush** or a wire wheel on an angle grinder if more aggression is required. If the stock on a butt joint is thick, the edges should be beveled to achieve better penetration.

Just as with stick welding, you need to clamp a ground wire (the negative side of the circuit) to the workpiece. The filler material (which is a spool of wire stored on the welder) acts as the electrode and should poke out the tip of the welding gun about the length of your little fingernail.

Hold the gun vertically, angled 10° away from the direction you'll be moving the welder, so that the tip will be driving forward toward the puddle. Practice the weld before initiating the arc, going through the full range of motion.

Then, with your welding mask down, strike an arc with the wire to start a puddle of molten metal. The filler wire will automatically feed into the puddle at whatever feed rate was selected. Drive the tip forward, with a slight weave back and forth. The angle of attack, along with the power and feed rate, may have to be changed depending on whether you're welding a lap joint, making a T-joint, welding upside down, or what have you.

Given its extreme ease of use, with practice a MIG welder can start to feel like a metalworking hot glue gun and is generally the most approachable for beginners.

TIG Welders

TIG stands for "tungsten inert gas." Invented in the 1940s, this is the premier welding technique, and it can handle just about any material and produce beautiful beads, making it the preferred style for highly visible welds, like those seen on bike frames. **TIG welders** are capable of welding steel, stainless steel, aluminum, copper, brass, titanium, and more.

TIG welders produce electrical current at the gun's tip and emit argon or helium gas that shields the welding process. A foot pedal allows the user to adjust the electrical current during the weld. The welding gun goes in the dominant hand, the nondominant hand holds the filler material, and one foot goes on the pedal. It's a complicated art and one generally reserved for those who have already mastered other types of welding.

To use a TIG welder, you first need to prep the surface metal, just as you would with a MIG welder. You'll also need to choose the correct gas and power. The tungsten electrode tip that sits within the gun must be ground to a fine point. There are charts that specify the ideal tip shape for particular materials. Angle the tip about 15° away from the direction you'll be traveling, and once you start welding, move the tip along the weld path just above the surface, feeding filler into it as you move along. All the while you must be careful not to touch the filler to the electrode; feed the filler into the puddle only. That's the gist, anyway.

With other welding techniques, a basic primer is likely enough to get you started on a self-taught course. But to properly learn TIG welding, hands-on instruction is essential.

Wrenches

A wrench is a hand tool that uses leverage to rotate fasteners like nuts, bolts, and other hardware (though the term *wrench* is also used to describe a tool that turns other objects, as in the case of the pipe wrench and the tap wrench). Wrenches and fasteners go hand in hand, and for every fastener that exists, a wrench was likely invented to fasten it.

Given their ubiquity, it's no surprise that wrenches have weaseled their way into our idioms. All manner of work can be reduced to "wrenching" on something. And the notion that one has "thrown a wrench into the works" as an act of impediment is universally understood, even among those who haven't spent time on the factory floor. In Britain, where wrenches aren't wrenches at all but are called "spanners," the saying is similar—one "throws a spanner in the works." It's a time-tested idea: The sturdy tools we all keep at hand to maintain equipment could just as easily foul a system. This is the plot of the novel *The Monkey Wrench Gang*, as well as a key takeaway from government produced handbooks on sabotage.

So it's not just an idiom; it's a real-world problem. Misguided wrenches have literally brought down aircraft, ruined spacewalks, and blown up nuclear missiles. Suffice to say, keep a tight grip on your wrenches.

Fixed Wrenches

A **fixed wrench** (or a **fixed spanner**, depending on which side of the Atlantic you inhabit) is a wrench that is nonadjustable. These come in a few shapes.

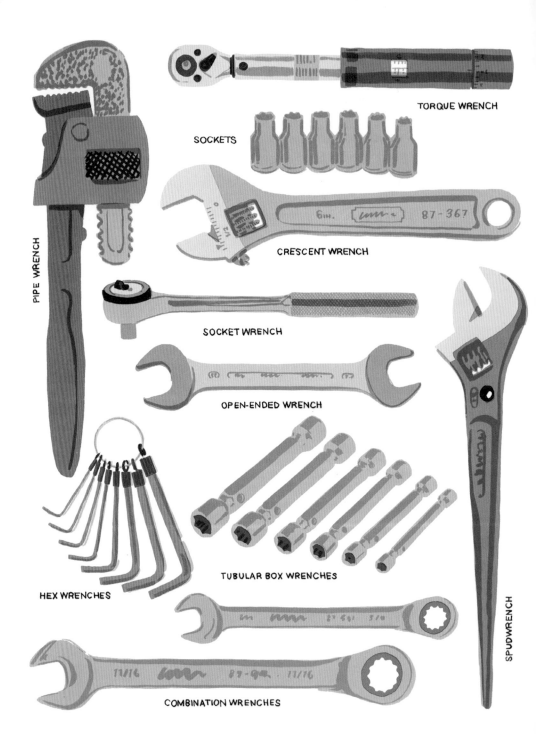

TORQUE WRENCH

SOCKETS

PIPE WRENCH

CRESCENT WRENCH

SOCKET WRENCH

OPEN-ENDED WRENCH

HEX WRENCHES

TUBULAR BOX WRENCHES

SPUDWRENCH

COMBINATION WRENCHES

For starters, there are **open-ended wrenches** and **box wrenches**. Open-ended wrenches, shaped like a crab claw, are meant to slip around a hexagonal nut or bolt. Box wrenches are similar, but they have a fully enclosed ring, generally rimmed with twelve notches. Besides giving more purchase by fully encircling the hex head, this system of twelve points means that the wrench can slip over a bolt in twice as many positions as the open-ended style. Turning a bolt in tight quarters, such as in an engine bay, means that you need only move the box wrench slightly to reset for another crank.

Both open-ended and box wrenches will almost always have a wrench head on both sides, of two different sizes, but many prefer a set of **combination wrenches**, which have an open-ended wrench on one side and the same size box wrench on the other. Kits come in SAE (standard, or imperial) or metric, and you'd do well not to confuse them, as hex heads are easily stripped by wrenches that almost fit but don't quite. Though if you find yourself in an emergency situation—broken down in some remote locale and forced to make an oversize wrench work—try shimming the gap with a coin or two.

Another style of fixed wrench is a **hex wrench**. Yes, the previous wrenches are used to turn hexagon-shaped nuts and bolts, but hex wrenches are used to turn fasteners that have a hexagonal *socket*. These are often the machine screws holding together bicycle parts, motorcycles, and self-assembled furniture. You've probably heard these wrenches referred to as **Allen wrenches** or **Allen keys**. Allen is a brand name, and even though the Allen Manufacturing Company released its wrenches in 1943—over thirty years *after* the invention of the hex wrench—their name nonetheless became synonymous with the tool. Other companies should be so lucky.

Hex wrenches often come on key rings or in foldout packages reminiscent of a Swiss Army knife. They're also available as individual tools in index kits, and these L-shaped wrenches often feature a ball end that allows the wrench to be used at an angle. For high-torque applications, or where frequent use is expected and comfort is prioritized over portability, larger **T-handle hex wrenches** are the way to go.

It's also important to note that Torx (or star drive) fasteners have become popular for similar applications, and the **Torx wrenches** for this drive are available in all of the same forms as hex wrenches, which can be confusing if you're not looking out for it. Using a Torx wrench in a hex bolt *might* work, but it's just as likely to strip the bolt. Also, just as with open-ended wrenches, confusing metric and SAE sizes with hex or Torx wrenches will mangle a hexagonal hole into a round one and really throw a wrench into the works, as it were.

Adjustable Wrenches

The most commonly used adjustable wrench is the **crescent wrench**. Crescent wrenches come in a few sizes, with different lengths and jaw spans. They look like open-ended wrenches but with a threaded barrel that's manipulated by a thumb to adjust the jaw opening. Normally, their handle is just a handle, but there's a variation known as the **spud wrench** in which the handle terminates in a metal spike. These are used to bring bolt holes into alignment without risking a finger.

With crescent wrenches, if you're unable to crank a full rotation and will need to take the wrench on and off to make partial rotations, try keeping a thumb on the adjustment barrel to quickly open and close the jaws some small amount. This way it will be easier to tighten the jaws on a bolt head, turn the wrench, loosen the jaws a touch so it can slip off, and then reset it in a new position to repeat the process.

A good portion of the planet also attributes the crescent wrench to Swedish inventor Johan Petter Johansson. (There's some global debate as to whether he was the first, as there seems to have been a lot of parallel wrench inventing going on in the late 1800s.) And because of Johansson's contribution, in Denmark the adjustable spanner is called a Swedish key, and in Russia it is called the little Swede. In fact this wasn't his only wrench. In 1888 he also invented the pipe wrench.

Monkey wrenches are similar and were one of the earliest adjustable wrenches. They feature smooth parallel jaws that are brought together with the rotation of a knurled barrel. But they're hardly used anymore, having been replaced mostly by crescent wrenches. Nowadays, if someone asks for a monkey wrench, they are likely asking for a **pipe wrench**—a similar enough looking bulky wrench with serrated jaws.

Pipe wrenches are one of the few wrenches designed to grasp onto round objects like pipes and large bolts. They tend to be hulking instruments, generally with cast-iron but sometimes aluminum frames. They have jaws with opposing serrations, and one jaw moves out with a turn of the barrel. If it feels sloppy, this is by design. That wiggle enables the wrench to latch on, but it only works in one direction.

The most important thing to know about using a pipe wrench is that it can only be used when the handle is rotated in the direction of the open jaws, not away from them. To use a pipe wrench, you tighten the jaws around the pipe and then apply pressure on the handle. The loose jaw will wedge tighter into the pipe, and the

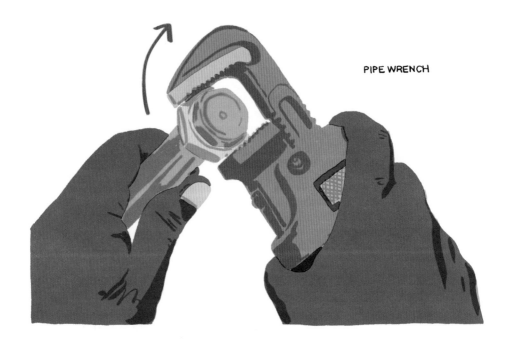

PIPE WRENCH

hardened serrations will dig into (and chew up) the comparatively softer pipe.

To grip a pipe and turn it without marring the surface, you'll want a **strap wrench**. The loops on these are generally adjustable to sizes larger than a pipe wrench can accommodate and will grip without biting. (You may see similarly styled wrenches used to swap oil filters on cars.) Like pipe wrenches, they rely on a cam-ming action to keep tension and can only be turned in one direc-tion, but they tend to at least do you the courtesy of printing an arrow on the wrench to help things along. Their downside is that their handles are usually shorter, and so leverage is lacking. A tradi-tional pipe wrench is often the more utilitarian tool of choice until a job demands the softer grasp of a strap wrench.

Socket Wrenches

In theory, a **socket wrench** should be any wrench that uses a socket fitting to cap over the top of a fastener and rotate it. The **lug nut wrench** in your car could be called a socket wrench, along with the **nut drivers** found alongside other screwdrivers . . . but neither are. In practice, the term *socket wrench* is reserved for a handle that affixes to detachable sockets—and that handle almost always has a ratcheting mechanism. You'll also see these called **ratchet wrenches** or just plain **ratchets**.

The design of socket wrenches has had an impressive run. The first ratcheting socket wrench was invented way back in 1863, and 150 years later, a digital version of a remarkably identical wrench was emailed to the International Space Station to be fabricated with a **3D-printer** *and put to use.*

Socket wrenches are almost always purchased as kits that include a wrench, a few adapters, and sockets in an array of metric and SAE sizes. The sockets—laid out in a tray like game pieces—have a square hole on one side to fit onto the wrench, and a six- or twelve-notch drive on the other. The square side snaps into the socket wrench, often with a spring-loaded ball detent (catch) that needs to be released with a push button. The driver side slips over a hex-head bolt or nut (though there are sockets that fit Spline drive fasteners, as well as internal and external Torx fasteners). Because sockets are fully capped, it can be a problem fitting them over the top of longer bolts, spark plugs, and anything else that extends longer than the socket is deep. **Deep well sockets** address this, though another solution is **ratcheting box wrenches**, which look like normal box wrenches but ratchet just as a socket wrench does.

Ratcheting wrenches are particularly popular with mechanics who work in tight confines, and socket wrenches with pivoting heads, swivel adapters, and extensions are even more adept, making these tools essential to any shop. Plus, the sockets themselves aren't limited to use on a socket wrench. With adapters, they can be fitted to a screwdriver, a drill, or an impact driver.

One of the risks with socket wrenches is losing whatever was housed in the socket—or losing the sockets themselves. Plenty of automobile mechanics have inserted a bolt into a socket and attempted to crane it into position, only to have the bolt slip out and drop into some inconvenient location. A trick to mitigating this risk is to put a piece of tape or shop rag into the socket and press the nut or bolt in behind it, but then there's the potential to lose the sockets themselves. Sometimes those ball detents on adapters fail or were never engaged to begin with, or the socket is just too heavy to remain in place. In some situations, you may want to wrap some electrical tape around sockets and adapters to avoid the risk of catastrophic mishap.

> How catastrophic can losing a socket really be?
>
> In 1980, during the Cold War—a decidedly inopportune time to accidentally detonate a nuclear missile—a twenty-one-year-old missile technician named Dave Powell dropped the socket from his socket wrench while working in a silo in Arkansas. The socket fell eight stories, puncturing the side of a Titan II missile that was fitted with a thermonuclear warhead. Rocket fuel was released, which caused an unfortunately deadly, but thankfully nonnuclear, explosion.
>
> The risks in wrenching over the open engine of your 1999 Volkswagen aren't quite the same as wrenching over a Titan missile, but then if it can happen in such a high-stakes work environment, how easily could it happen in your garage? It's worth saying again: Keep a hand on those wrenches.

Torque Wrenches

Torque wrenches are tools that measure the amount of rotational force being applied to a fastener. For a low-stakes application (like assembling a desk), this isn't necessary, but in construction, on machinery, and in the automotive and aeronautical industries, torque values are specified, and a failure to abide by these specifications can result in disaster. Consider Emery Worldwide Flight 17, a cargo plane that *was* headed from Reno, Nevada, to Dayton, Ohio. It crashed into a Sacramento salvage yard, all because bolts on the elevator control rod were torqued to an incorrect value.

Torque wrenches come in four forms. The first is a **deflecting-beam torque wrench**, which, as the bolt is tightened and the wrench torqued, moves a needle across a gauge like the gas gauge of a car. The second is similar, but it uses a round dial. The third is a **digital torque wrench**, and these tend to be pricey. Lastly, there's the more common **click torque wrench**, where the torque value is set by rotating the handle to the desired value. As the wrench is torqued, you'll know you've tightened it to the proper amount when a click is both felt and heard. These click-style wrenches should be stored at their lowest setting, so as not to compress the internal spring for long periods.

In the United States and a few other countries, torque values are expressed in foot pounds or inch pounds. (Technically, the true unit of measurement for torque is the *pound-foot*, but it's a technicality that no one ever abides by, and the order is always reversed.) This value is exactly what it sounds like: the amount of rotational force applied when one pound of force is exerted at the end of a 1-foot lever. In most other countries, torque values are expressed in newton meters—a newton of force on the end of a 1-meter lever.

To give all this some context, consider that the lug nuts that hold your car's wheels in place are probably tightened to about 80 foot pounds [108.5 Nm], a torque value that would be achieved if your mechanic stood his 80-pound niece on the end of a 1-foot-long socket wrench.

Torque is the measure of force used to rotate a nut or bolt, but it's actually a means to an end, because what you're really aiming to do is create a certain amount of clamping tension between nut and bolt. When a threaded connection between two points is torqued, the bolt actually stretches and becomes a spring that pulls the two surfaces together. This clamping force distributes the load from the bolt to the materials being pinned together, and it also helps prevent the threads from turning and the connection loosening.

Because bolts are stretched in these conditions, some applications or manufacturers will specify that bolts can't be reused. Putting the head of a car engine on, for example, requires torquing bolts in a specific pattern to an exact setting. Many mechanics, respecting either the manufacturer's request or their own paranoia, won't reuse previously stressed bolts when performing this work.

Torque wrenches aren't the only tool that can achieve a specified torque value. **Torque screwdrivers**, as well as programmable cordless drills and impact drivers, can be set to a desired torque. And **torque multipliers**, many of which look similar to a cordless drill but use gear reduction to generate immensely more torque, are capable of mounting propellers to warships and wind turbines to their foundations.

How much torque are we talking about here? One model of **hydraulic torque multiplier** is supposedly capable of turning a bolt with an incomprehensible 58,000 foot pounds [78,600 Nm] of force. To revisit the example of your car's lug nuts, this is the absolutely immense amount of torque that would be generated if your mechanic somehow hung six fully grown elephants from the end of his 1-foot-long wrench.

SHAPE

Angle Grinders

Angle grinders are handheld power tools that spin an abrasive disc at high speed, generally for the purpose of grinding away, cutting, or shaping metal. It's a no-frills device, with just enough handle to contain the rotating spindle (called the *arbor*), a motor, and a switch. The handle is just big enough for one hand, and a second accessory handle is threaded into the body at a right angle. A removable and adjustable guard covers about half of the disc. Some angle grinders have trigger locks, which might seem useful but mostly exist to injure those who plug in a trigger-locked grinder. Angle grinders come in a few sizes and are sold according to their maximum disc diameter—generally 4½ to 9 inches (or 115 to 230mm in metric regions).

The great thing about angle grinders is that their utilitarian design means they're ripe for adaptation, so they're used for all kinds of tasks beyond metal grinding. Metal **cutoff wheels** are the most popular. An angle grinder outfitted with one will shower a lot of sparks, but with the right disc there's no metal you can't cut through. It's a necessary tool for cracking into a safe—and probably for breaking into the building containing it. They can even be used to slice synthetic rope, offering the added benefit of melting the rope ends to keep them from fraying. There are also **wire brushes** that quickly remove rust. **Flap discs** of layered sandpaper can also be used for rust removal or for sanding wood. **Chainsaw discs, planing**

discs, and **rasps** all hog out wood in a way that smaller rotary tools or chisels just can't match. **Scraping wheels**, like those made by Diamabrush, strip mastic and finish, reducing flooring to bare wood or concrete. One can even attach **diamond blades** for stone and tile. And then there are even stranger attachments: actual chainsaw blades about the length of your forearm. **Belt sanders**. Power chisels. Drill chucks that convert your angle grinder into a right-angled drill. It turns out that when you offer a tool that's little more than a motor on a stick, people will find a lot of ways to put it to use.

The most important thing to consider with angle grinders is safety. This tool spins faster than almost anything in the shop, some up to 12,000 rotations per minute (rpm). That's four times the speed of a lawn mower blade. And those discs aren't infrangible. One crack, or some pressure at the wrong angle, can cause them to shatter, sending jagged shards in all directions. Disc failures and unprotected faces are a sobering mix.

Always wear safety glasses *and* a face shield when using an angle grinder. Loose clothes can get wrapped up in the disc, so be mindful there. If you're cutting or grinding metal, wear natural fabric that can handle the sparks. The thick stream of sparks thrown from metal grinding can be nearly 2,000°F [1,000°C], so keep combustibles clear. Use two hands on the tool, and be sure to keep the adjustable guard between you and the disc.

Grinders spin at different speeds, which you'll find indicated on your grinder. The disc's maximum rpm is also printed on the disc. Never use a disc with a maximum rpm lower than the grinder's maximum speed. Centrifugal force quadruples when rotational speed doubles, which means a disc can shatter far more easily than you might assume, giving you a chance to find out which soup the hospital is serving for dinner that night.

Use the correct disc for the job. This goes beyond the broad strokes of using wood-carving discs for wood and metal-cutting discs for metal. For example, a disc meant for steel will tear up aluminum, and a disc that's been used on aluminum shouldn't be used on steel.

If you use a disc on steel that's been used on aluminum, you will create a crude version of thermate. What's thermate? Thermate is the main ingredient in many of the military's incendiary grenades. So be sure you use your disc only on the metal it's labeled for.

There are a lot of ways to thread on the various attachments and discs, so be sure to read their instructions. But in general, the discs slide onto the threaded arbor with the label facing the body of the grinder. The nut that locks down the disc has an offset edge that faces the disc on thick grinding wheels and away from thin cutoff discs. Large wire wheels and their ilk might have their own built-in nut that threads onto the arbor. For these, you wouldn't use the backing washer or nut. Regardless of the method of securing the disc, before powering up the angle grinder, make sure the disc is flat, secure, and rated for the grinder and that the guard is a good fit.

Chisels

Chisels are a forward-facing blade with a handle that may (or may not) be designed to be struck by a hammer. This basic functionality means that they're as much a trade-spanning tool as the hammers used to strike them. Stoneworkers, plumbers, mechanics, woodworkers, leatherworkers, and electricians all use chisels.

You'll find general-purpose chisels at your hardware store. These have cheap steel and a plastic handle with a length and build that doesn't really call out for any specific purpose. They're often relegated to living a life of abuse at the bottom of a toolbox, rarely sharpened, and used to pry open paint cans. Those tools have their place and may even be exactly what you're looking for to accomplish a one-off task. But for woodworking, carpentry projects that require precision, metal fabrication, and masonry, there are purpose-built chisels that you should understand and care for.

Wood Chisels

There are more types of woodworking chisels than can possibly be covered in this book. Part of the challenge is that chisels can be so visually similar that their subtleties are difficult to appreciate in text, but just as problematic is that each trade and locale has adopted its own terminology, which of course evolved from generation to generation. It's a knot that can't easily be untangled, so we'll reduce chisels to a few agreed-upon categories.

Let's start with **bench chisels**. As the name implies, these are the chisels that sit atop a bench and are frequently called to task for general-purpose woodworking. They have a blade the length of a hand or a bit shorter, with beveled sides and a flat back. **Butt chisels**, often thrown into the bench chisel mix, are similar in build, if a bit wider and shorter and with a larger pommel to tuck into the palm.

Stouter chisels are called **mortise chisels**. These are square sided and beefed up, meant for the pounding and prying task of excavating the mortise of a mortise-and-tenon connection. (These are, respectively, the interlocking recesses and protrusions of wood joinery, and their layout can be marked with a **mortise gauge**.) **Corner chisels**, as their name would suggest, are used for cutting right-angle corners in those mortises.

Paring chisels are longer and more delicate. They're meant for the hand-powered work of slicing off thin bits of grain (a process called *paring*); you wouldn't use a mallet to strike a paring chisel. These chisels are further divided by the angle of attack they offer. For example, **left- and right-skew chisels** have angled blades for touching up opposing faces of dovetail joints, and **crane-necked chisels** have an offset handle, allowing the full flat bottom of the chisel to act as a hand plane.

Then there are **framing chisels**, which are used in *timber framing*, the infrequently practiced art of using large joinery and girthy beams to frame buildings without mechanical fasteners. These chisels are larger than the others we've discussed, but they have all the familiar features and are robust enough to strike with a mallet. There are flat-bottomed framing chisels for more delicate paring.

Corner framing chisels for their obvious application. And wide framing chisels, called **framing slicks**, for planing greater surface area at once.

Generally speaking, chisels are not meant to remove a lot of wood. The exception to this is **turning chisels**, which are braced on a **lathe** and used to turn large hunks of timber into bowls and balusters and such. The use of turning chisels can get pretty aggressive, and they're built for it, but with all other chisels, a more measured approach is appropriate.

It looks something like this: Unless the particular chisel is prohibitively large, hold the chisel in your dominant hand like a pencil, positioning it so that the blade will take just a small bite of grain, and strike the back of the chisel with a mallet. Or use your nondominant hand as a guide along the blade so that your fingers act as a depth stop, and push the back of the handle with your dominant hand. Both techniques are in service of taking off small, controlled shavings of wood. Pushing into too much wood will cause the chisel to bind, and if you do muscle through, the chisel will drift and you'll lose accuracy.

A good general approach is to saw or drill most of what can be sawed or drilled, and only then finish to your line by chisel. For best results, pare by hand half of what wood remains (or use the mallet to strike the chisel, if appropriate), then half again, and half again until one final pass through a wispy section of grain is all that's left.

A final note: A wood chisel needs to be sharp. A sharp chisel allows for a carefully controlled cut. A dull one will drift, mangling the wood grain and eventually mangling you. You can read about how to keep your chisel sharp on page 96.

Cold Chisels

Cold chisels are metalworking chisels; they're the room-temperature counterpart to the **hot chisels** used by blacksmiths. With a cold chisel, unheated metals can be bent, shaped, and cut. The four basic shapes of cold chisel at your disposal are the flat chisel, cape chisel, half-round chisel, and diamond-point chisel.

The **flat chisel** is the most frequently used. A common use is for shaving off rivets or breaking off the heads of stripped screws and seized bolts. A flat cold chisel can be used to cut sheet metal, and a lighter tap of the chisel can crease the metal as a **sheet metal brake** would, so a crisp bend can be formed.

Cape chisels, which are spear shaped with a thin cutting edge and narrow point, focus a lot of force into a tiny area, and can be used for cutting narrow grooves. Cape chisels can pry into the tight gaps of machinery to extract oil seals, or they might be used for *staking* a nut—intentionally deforming a nut or a bolt's threads to foul the path of a nut, locking it in place.

At first glance, the **half-round chisel** (sometimes called a **round-nose chisel** or **gouge chisel**) seems to have a diamond point, but the tip is actually blunted, and one side is rounded over. This shape allows the chisel to cut channels with round bottoms. Round-bottom grooves might be decorative, or they could be ball-bearing channels or oil ways for lubrication.

In a similar vein, there's the **diamond-point chisel**. This stout point is used to cut metal or carve V-shaped channels, be they functional or decorative.

Masonry Chisels

Masonry chisels go by a few names: **concrete chisels**, **brick chisels**, and **bolster chisels**. The terms *brick* and *bolster* tend to be reserved for wider models, but just as often, all these names are conflated. The important thing to remember is that these chisels are meant for cutting and shaping brick, concrete, and stone, while similar-looking cold chisels are not. Before you use a chisel to crack some rock, take care to ensure that it's designed for the task.

To use these chisels for cutting masonry, you'll need a **club hammer**. Place the chisel along the line where you want to cut the brick or block, and strike it once or twice with the hammer. Then move down the path and repeat the process until you've scored a line all the way around. Keep working around and deeper until the block fractures, hopefully along the path you just provided. Any knobby remnants can be cleaned up with further chiseling.

There are other options. For a quick and dirty cleave, a very wide brick chisel can break a brick more or less where you strike it with a single blow. For a cleaner line, you could use an angle grinder or a circular saw with a diamond blade. Either would be capable of eventually cutting the brick, but you can also use these tools to score a crisp line on the visible face, and then split the remainder of the brick with a chisel. For larger jobs, you can rent a **block splitter** or a **brick saw**, which is essentially an oversize wet tile saw. Though you'll still want a chisel handy for the odd nub that needs tending to.

Drawknives and Spokeshaves

A **drawknife** (or **drawing knife**, as it was once called) is a knife blade with a handle on each end; it's meant to be pulled (or *drawn*) toward the user. That might sound dangerous, but between the control of two hands safely beside the blade and the blade itself being recessed beyond the handles, it's thankfully less likely to injure than you might imagine.

Drawknives have either curved blades or straight blades, and some handles can collapse and change position, though the handles on most tend to be fixed and pointed straight at the user. Deeply curved drawknives, called **inshaves**, are used to hollow out bowls and chairs, not unlike how you'd use a scorp.

With drawknives, the blade length puts your hands about steering wheel distance apart—a wide enough grip that allows shoulders and elbows to draw back in a full stroke reminiscent of rowing a boat. Ideally, this is done on a **shaving horse**—a foot-powered clamp-and-bench combo that holds a piece of timber in place so it can be shaved down.

With a sharp blade and the aid of a shaving horse, a drawknife is highly efficient at removing wood. During World War II, fifty million British tent pegs were manufactured with drawknives, with the goal of using only eighteen draws to reduce a log down to a tent peg.

This efficiency might also be used to rough out cylindrical billets for the wood lathe, or for debarking entire tree trunks. For more delicate tasks like cutting a crisp bevel or smoothing out facets on spindly wooden members, there's the **spokeshave**.

A spokeshave is similar to a drawknife in that it's gripped with two hands and drawn toward you. However, the handles of a spokeshave point to the left and right; and rather than having a large exposed blade, the blade is neatly housed.

Spokeshaves are used in much the same way as drawknives, but rather than cutting deeply and removing large amounts of wood, they pull off measured ribbons with control. Traditionally, they were used for making wagon wheel spokes (hence the name), but they might also be used to shape the legs and spindles of chairs.

Like drawknives, spokeshaves have many cousins. **Chairmaker's spokeshaves** have a convex blade for hollowing out the seats of chairs. In traditional chair making, this is also a job performed by a tool called a **travisher**. Any spokeshave can be used to put a chamfer on a corner, but **chamfer spokeshaves** are specifically meant for the task; they have a notch that rides along an edge. There are also **beading tools** that look just like spokeshaves, but their blades are shaped to cut a variety of decorative profiles as they're drawn along the workpiece.

Then there's a spokeshave that's not recognizable as a spokeshave at all: the **tenon cutter**, also known as a **dowel pointer** or a **spoke pointer**. The tenon cutter is mounted within a cone and attached to a drill bit shank or clamped in place. In essence, it's a very large pencil sharpener for wooden dowels or perhaps for the chair spindle you just shaved down with a spokeshave. As the cone spins, the end of the dowel is tapered until it can tightly slot into the corresponding hole in the chair's seat. You could accomplish this taper with a drawknife or spokeshave, but the spoke pointer will give a more consistent taper. And if it's an untapered dowel you're looking to form, versions of those tenon cutters exist as well.

Like all bladed tools, drawknives and spokeshaves need to be kept sharp. The curved blades of drawknives can be a challenge. The puck-shaped sharpening stones used in ax sharpening are well suited for the task. Leather strops also conform to the shape easily, and keeping up with stropping is a good habit that will stave off more involved sharpening. But when the time does come, follow the sharpening procedure outlined in the sharpening section (see page 93).

USING A DRAWKNIFE ON A SHAVING HORSE

Files

The first files were abrasive stones. Then came the soft metal files of the Bronze Age. And finally, we arrived at the modern steel file. While the basic modern design has not changed much since, originally these files were hand cut and labor intensive to produce. Leonardo da Vinci designed a machine to automate the process, but like many of his creations it proved to be only theoretical. A few hundred years later, a French locksmith created a functioning design, and by 1800, machine-cut files were being manufactured. But as it happened, as soon as files could be churned out, the demand for them dropped swiftly. Ironically, while automation was producing files en masse, it was simultaneously erasing the need for them.

Why? Files used to be a critical part of all manufacturing. Most products were built from start to finish by a handful of artisans, and individual components were fine-tuned (with files) to work within a finished product. This all changed in the late 1800s as the world pivoted to manufacturing that produced interchangeable parts and an assembly line that allowed a worker to focus on only one slice of the process. Interchangeability is a concept that's taken for granted now, but at the time it was so revolutionary that when several American manufacturers demonstrated their products' interchangeable components (namely, in firearms) at London's Great Exhibition in 1851, the system was dubbed the *American System*, and it was the talk of the event. The world never looked back, and manufacturing moved from artisanal to cookie-cutter.

Files are lengths of hardened steel with parallel rows of cutting edges used to remove material from metal, hardwood, plastics, and even stone. They remain essential to much of fabrication work, where they are used to clean the burr from freshly cut metal and to make minor adjustments to fit. As the exuberance of grinders can generate too much heat, files are also called upon for blade and tool care; any owner of a lawn mower, shovel, or saw should be using a file to touch up those edges.

There are a few ways to organize files, but the easiest is to first divide them into two categories: **American pattern files** and **Swiss pattern files**. American pattern files are available in six types of cuts, from coarsest to smoothest: rough, middle, bastard, second cut, smooth, and dead smooth. Swiss pattern files are offered in seven cuts, ranging from the coarsest, OO, to the smoothest, 6. Swiss pattern files are smaller and finer than American pattern files and are generally reserved for detailed work, such as model making, tool and die work, and jewelry. Within those two broad categories and grades of cut are a couple of other distinctions. A file's face can be single-cut or double-cut. A single-cut file has one pattern of parallel teeth, while a double-cut file (also known as a *crosscut file*) has a second pattern of cuts forming diamond-shaped cutting edges. Files are also categorized by the shape of their cross section: flat, tapered, round, half round, and triangular.

There are a few other alternate names for the file shapes and cuts already covered. **Mill files** and **flat files** are nearly identical and probably what you imagine when you picture a standard-issue file. **Machinist's files** are similar, though often double-cut. **Hand file** is a term that could encompass every type of file out there, but the name is often used for a flat, general-purpose metalworking file. **Needle files**, miniature versions of all of the above, are relegated to detail work. **Diamond files** are in a category of their own; they're impregnated with ultra-hard diamonds and are used for glass and stonework. Then there are the files named for a specific job. **Saw files** are meant to sharpen the teeth of a saw but are also used anywhere a smooth finish is desired. **Chainsaw files** sharpen teeth as well but are also called upon to remove height from the chain to establish cutting depth. And **fret files** are meant just for rounding over the metal ribs that texture the neck of a guitar.

The use of files is fairly straightforward. The thin pointed end is the *tang*, and it's meant for sliding into a handle, though one isn't required; you're welcome to just grip the file by the tang. Files cut on the push stroke only, so you'll need to hold the file diagonally to the direction it will be moved, flat on the surface of the workpiece, and push it along. Be sure to move the file across its whole length so that all of the cutting edges are used. Then lift it up, bring it back, and repeat.

Draw filing is another filing method that can offer more control. This involves holding the file at both the handle and the point with both hands, and pulling it diagonally across the workpiece toward your body, like a drawknife.

Chalking a file can help keep it from becoming loaded with metal shavings. The teeth can also be cleaned out with a **file card**—an oddly chosen name given that this tool very clearly looks like a brush. Files should be thoroughly cleaned between uses, and this is all the more crucial when transitioning between materials.

Gouges

Gouges are concave chisels used in wood carving and wood turning. Like chisels (and they are often categorized as a subset of chisels), they exist in more sizes and shapes than can easily be cataloged here.

They're generally first organized by whether they're intended for wood carving or for turning wood on a lathe. Then they're grouped by size and sweep, the *sweep* being the degree of that quintessential U-shaped curve common to most gouges. A gouge's sweep is defined as flat, middle, or regular, though there are some gouges that have a sweep so extreme that they're called **V gouges**. Gouges can also be organized by their varying tapers (or lack thereof), which you'd identify by looking down at the overall outline of the tool's blade and shaft. **Straight gouges** have parallel edges, while **fishtail gouges** flare out to points that are excellent for working into corners.

With the common U-shaped gouges, minute changes in the bevel and sweep will produce different cuts, but don't let the options overwhelm you. A few of gouges is plenty to outfit a wood-carving kit. Of course, as is often the case, a dedicated wood-carver who appreciates the subtleties of each might collect a vast collection of antique and modern gouges—a collection that morphs from "tools I need" to "tools I appreciate" and becomes the envy of fellow woodworkers and the bane of romantic cohabitants.

The care of these tools is not unlike the care required for the hooked blade of a spoon knife. (In fact, if you've got a vise (or clamp) but no hook knife, a gouge is the perfect substitute for hollowing out the depression on a wooden spoon or bowl.) The curved blade of a gouge is a real pain to sharpen, a process that's more difficult the smaller the blade of the gouge. (The **vein gouge**, named for its purpose of etching thin lines onto carvings of foliage, has a curved blade so small you'll wonder how much a new one costs while sharpening your old one.) So, it's best to keep a strop handy and to keep delicate blades away from hard surfaces.

Strop the blade after each use. When the time comes for a true sharpening, you have a few options. You can use sandpaper on a soft surface, like a rubber mat. Roll the blade as you move it across the sandpaper, following the curve of the gouge. The underlying rubber will deform some around the curve and help keep flat spots from showing. If you don't have a rubber mat, follow the usual guidelines for sharpening (see page 93), but expect a lot of practice before you master a smooth motion around the belly of the gouge.

Jointers

A **jointer** is a power tool used for creating that first flat face on an otherwise unflat piece of wood. What makes this possible is the tool's lengthy table, which houses a recessed cutterhead that's flush with the table's surface. As wood is slid along the table and over the rotating blades, the high spots are shaved off, pass by pass, until one final pass across the entire face produces that flat plane.

Jointers differ from many other woodworking tools in that they cut the same face they reference. Consider **thickness planers**, which cut the opposite face from the one they reference. This is how they're able to shape boards of consistent thickness, but this means if a board has a bow to it, it will come out the backside of a thickness planer with that same bow. (If you're wondering why you

The tool owes its name to the *joining* of wood. Seamless connections require flat faces and square timbers, so something like the boards that make up a tabletop will have been edge jointed before being glued together. Perhaps confusingly, **joiners** are a separate tool altogether and yet also have their place in building that tabletop. These tools are used in furniture making to cut precise slots and mortises for pegs and other manufactured tenons (such as biscuits and Dominos). In theory, you might use a **biscuit *joiner*** and its torpedo-shaped tenons to create solid mechanical connections between the tabletop's butt joints, which of course you had already edge-*jointed*. Get it?

can't just use a joiner on both sides of a piece of wood in place of a planer, it's because a joiner will produce two flat faces, but not two *parallel* flat faces. So you might end up with a wedge-shaped piece of wood.) Likewise, a board can be run through a table saw and a fresh edge cut, but that cut will only be straight if the opposite edge that was riding along the saw's fence was straight. For the planer to produce a board that has parallel and flat sides, and for a table saw to produce a board that has flat and parallel edges, they both need to start with one side and edge that's truly flat. This is where the jointer comes in.

Except . . . you don't necessarily need a jointer for this job. Some woodworking shops certainly find them indispensable: Large jointers can flatten huge timbers with ease, and there's no better solution for truing up wood. But these bulky hunks of cast iron are often the last of the large tools purchased by woodworkers, and many forgo them entirely in favor of alternative techniques.

For starters, you can use a **hand plane**. Large planes, which are sometimes called **jointer planes**, are built the same way as a jointer: a long metal shoe and a blade housed in the middle. These planes are used in conjunction with **winding sticks** to flatten a board. The two sticks are placed along various spots and you eyeball how their lines compare. Their length is generally longer than what they're resting on, so they exaggerate the wavy bends of the board. Get down level with the sticks and compare them. The rear stick is a little high on the right? Plane the rear right corner of the board down, and check the sticks again.

You can use a jointer plane on both the face and the edge of boards, but if you want to save some time on the edges, use a

router with a **flush trim bit** and a clamped straight edge as a guide. In that same vein, a circular saw and guide can also joint an edge, or a track saw can be used for even better results.

Another option is making an **edge jointing table saw sled**. This device, which is like a crosscut sled with a couple of toggle clamps mounted to it, will hold a piece of lumber in place, allowing you to move it through the table saw blade on a straight path guided by the saw's miter slots. This is necessary because if you just push the lumber along the saw's fence with no jig, the cut will just follow whatever undulations are on the opposite face.

Then there's a **router sled**, which can be made from plywood or aluminum. This is the preferred method for flattening large slabs of wood that can't actually fit on most jointers. The sled is a trough for the router to sit in; the sled slides up and down rails that straddle the slab. This way the router, with its bit spinning at a consistent height, can move across the slab, leveling it inch by inch, like a typewriter carriage processing a piece of paper: back and forth, one line at a time.

A sled can be made for a thickness planer as well. The board being sent through is placed on a plywood carrier, and all the little gaps and raised corners on the bottom are shimmed and tacked in place with a **hot glue gun**. This keeps the rollers of the planer from compressing the board, which would otherwise spring back to its twisted or bowed shape. If the board is incompressible with shims, and the bottom of the sled is truly flat, then the drum of knives in the planer will have no choice but to cut a flat surface on top. No jointer required.

Lathes

Lathes are considered by many to be the mother of all machining tools. It's often said that a lathe is the only tool that can reproduce itself, which isn't entirely true, but that close-enough adage drives home a point: These are very capable machines.

Lathes—especially metal lathes—are immensely powerful. This power should be taken seriously. The lathe is easily one of the most dangerous tools in the shop (if not *the* most dangerous). While plenty of tools will tax you a digit or two, lathes are a crocodile lying in wait at the water's edge, ready to drag you under. Loose clothes or hair will quickly snag, and faster than seems possible a large lathe can pull you into its spindle and reduce your human shape to a tightly wound bundle of sinew. Likewise, the lathe's chuck key needs to be policed at all times. A chuck key flung from a rotating chuck is a good way to break a window or loosen some teeth.

Volumes of books have been written on the subject of turning—especially metal turning, given the tolerances and the risks with such a powerful tool—so any foray into this area should include a healthy amount of research and ideally a class at the local community college. There are also a growing number of membership-based workshops that provide both access to lathes and the requisite training to use them.

There are two fundamentally different types of lathes: machine-shop lathes meant for metal (though plastics and some wood are also fair game) and lathes used only in woodworking. Both work by rotating a workpiece around an axis so that it can be shaped (a technique known as *turning*), but exactly how this shaping happens varies. **Metal lathes** are for precision machining, with tools mounted on fixtures that can be adjusted by a thousandth of an inch or centimeter. **Wood-turning lathes** are less complicated and comparatively less accurate, and all of the tool work is handheld.

Metal Lathes

Metal lathes have more power than wood lathes, as quite a bit of torque is required to shave away steel with a tool edge. The tools of these lathes, or bits as they're often called, are mounted in place. **High-speed steel bits** are often ground to the correct shape by the user, but preshaped sets are available. **Carbide bits** are also an option. Tools are mounted in various tool holders affixed to the tool post, which rides on the cross slide, which in turn is mounted to the carriage. Each of these individual components can be adjusted, and it's their ability to be advanced in minute increments that makes this a precision machining tool.

The speed must be set before you start. Choosing the speed is no trivial matter, and it can't be fully covered here. Using a metal lathe requires an astute understanding of the cutting speed of a tool and the feed rate of a material—a complex bit of math based on several variables but truncated in machining parlance as "speeds and feeds." Every material, tool diameter, and cutting path comes into play to determine the speed of spindle rotation and the rate that a cutting tip is moved through the workpiece.

The piece of metal that is being machined is set into the lathe's chuck and tightened down, not unlike a drill. The bit is mounted, and the height of it is centered on the center of the workpiece. The very tip of the bit is the only part that does any cutting, so the bit must also be angled to give proper clearance to the noncutting edges. Most machining starts with *facing* the piece—that is, machining a flat surface across the front that's perpendicular to the axis of rotation.

To face the workpiece, the lathe is first powered on. Then the bit is slowly brought toward the rotating face by cranking the carriage controls until it just contacts the workpiece. This is called *touching off*, and you do this to find the high point so that you can cut just a little bit deeper. After touching off, the bit is brought off to the side of the workpiece and past the face some small amount. Then the bit is moved from the edge to dead center, cutting the mirror finish of a machined face as it passes.

Turning the workpiece along the length of its axis can happen in a lot of ways, but just as with facing, most machining starts with a first few passes to achieve *concentricity*, which is just a fancy word for cutting a fresh surface while making the object truly round. This process is the same, where the bit is touched off to figure out just how far you can cut in, and then the bit is driven into the shaft of the rotating workpiece by way of the carriage hand crank, leaving in its path a perfectly machined cylindrical surface.

Some lathes have a power feed to automatically advance the carriage at a set speed, and some can also synchronize with the rotation, which allows screw threads to be machined. There are also computer-controlled lathes (known as **CNC lathes**) that automate the entire process, and much like **CNC mills**, these dominate commercial manufacturing.

Wood Lathes

Wood lathes operate much the same as metal lathes in that they're large hunks of cast iron rotating a workpiece on a central axis, allowing that object to be shaped with tools. The technique is quite different, however, as is some of the nomenclature.

A wood lathe has a *headstock* that houses the *spindle*, which is the part that rotates the wood. Mounted to that spindle might be a *chuck*, a *spur drive* (also called a *spur center*), or a *faceplate*. Next is the *banjo*, and mounted to that is the *tool rest*; these slide along the middle of the lathe and are adjusted to suit the needs of each piece and cut. On the far side of the lathe is the *tailstock*, which has its own center (called a *quill*) for supporting the opposite end of the wood being turned. The quill is used when turning something long, like a chair leg, but it might be removed to free up space when turning the face of a stout object, like a bowl.

To turn wood on a lathe, you'll need to mount it to the spindle. This first means identifying the wood's middle so that it can be mounted dead center. On small square blanks of wood, you can draw lines from corner to corner and use their intersection as the center. But for round and irregular shaped pieces, you'll want what's called a **center-finder tool**.

The biggest difference between metal and wood lathe tools is that wood lathe tools are handheld. **Turning gouges** and turning chisels—which are stiffer than regular chisels—are held with two hands and supported on the tool rest. The banjo and tool rest are positioned to provide a perch for support and leverage.

A woodturner can go deep on tools. Every tool is capable of producing a slightly different profile, and many draw from a deep quiver. But to get started, only a few basics are required. A **spindle gouge** and a **skew chisel** offer curved and flat utilitarian blades to work with. A **parting tool** can be used to force a divide into a spindle, cutting a deep channel into and through the wood. A **bowl gouge** has a deeper cup than the spindle gouge and will get you started on hollowing out the interior of a bowl. And a **scraper** can be used for leveling out the tool marks left over from all the previous tools. With these and a little sandpaper, you're ready to turn.

Wood turning is physically demanding. The tool is mostly held with the back hand, which controls the movement. The front hand dampens the shock and helps guide the position and angle. You need a stable base with both feet firmly planted. Let's say you're going to use that spindle gouge to put some wavy profile onto a table leg. Practice the whole range a few times, moving the tool in a smooth motion, with elbows tucked in and gouge held steady. It will have to be a well-honed motion to resist the vibration once you're actually removing wood chips.

Lay the bevel of the gouge on the wood, and ease onto the point until you're throwing wispy shavings. As the wood rotates around, some grain will be easier to shave than others, so there will inevitably be some chatter. It's imperative to keep the gouge in place, in part to produce a clean and consistent shape, but mostly because if you catch the tip and the gouge is leveraged out of your hands, it won't be a pleasant experience.

After the desired profile has been cut, the table leg can be sanded on the lathe. Don't wrap the sandpaper all the way around the wood while it's spinning, or it will snag. Move a strip of sandpaper back and forth along the length of the wood, and then feel free to burnish it by repeating the same process with a handheld wad of shavings.

If you followed all the steps to this point, you should have a finished and completely useless single table leg.

TURNING ON A WOOD LATHE

Milling Machines

Milling machines (or, more simply, **mills**) are precision machining tools. Usually, they're used to mill metal, but they can machine plastics and even wood. With a rotating spindle suspended over a table, mills look like overstuffed drill presses. But rather than a meager apparatus that's only built to withstand the forces of being pressed straight down, the spindle of a mill is strong enough to resist the sideways force of a workpiece being fed into it.

The table of a mill is referred to as the *bed*. It might have a built-in vise and, depending on the complexity of the mill attachments, might even rotate. Handles are positioned on the sides and in front of the bed to maneuver it in several different directions, or *axes*. Side-to-side movement occurs along the x-axis; front-to-back movement along the y-axis; up-and-down movement along the z-axis. A crank called the *knee* is generally used to move the bed up and down, though some mills don't have a knee and instead raise and lower the spindle on an adjustable column called the *quill*.

All these directions of movement, and the ability to control them by amounts less than the thickness of a hair, make milling machines very capable tools in the right hands. There is almost no metal part in existence that can't be made on a mill or on a lathe, even if only a select few humans can figure out the manipulations required to do it. Machining is an art. The process to shape a block of metal stock can be a dance requiring many dozens of sequential steps—an

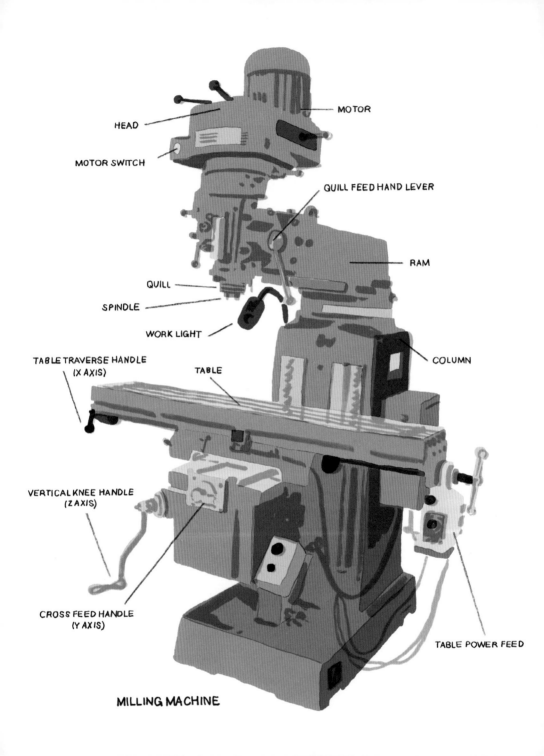

HEAD

MOTOR

MOTOR SWITCH

QUILL FEED HAND LEVER

RAM

QUILL

SPINDLE

WORK LIGHT

COLUMN

TABLE TRAVERSE HANDLE
(X AXIS)

TABLE

VERTICAL KNEE HANDLE
(Z AXIS)

CROSS FEED HANDLE
(Y AXIS)

TABLE POWER FEED

MILLING MACHINE

order of operation only discernible to those with years, or perhaps decades, of experience. Many elderly machinists are viewed as veritable wizards by their peers and their wisdom is still sought as they near a century in age.

There is an assortment of mills out there. Most are vertical, but some horizontal mills have the spindle mounted on its side. Sizes vary from micro versions to benchtop models to cast-iron behemoths that weigh a literal ton. But all mills are measured against Bridgeport mills. These are the gold standard of full-size mills, and those from the mid-twentieth century are highly sought-after. Many are retrofitted with motors or a digital readout (DRO), which displays the distance each axis is moving on a screen, but these machines are otherwise unchanged and, if well cared for, will likely remain in service for another century. A highly skilled machinist and vintage mill might have a combined century and a half of experience between them, and it's a prized pairing.

While there are a wide variety of manual mills and their operators, it shouldn't go unsaid that computer-controlled mills (known as CNC mills) are just as prevalent and have all but fully taken over manufacturing. CNC mills have evolved into full machining centers, often incorporating lathes and a few more axes of mobility. Many can swap their own cutting tools, manipulate a workpiece, and perform all this with a speed and dexterity that is beyond comprehension.

Here are the basics of how a manual mill works: A compatible tool is selected and fixed into the spindle, either in the appropriate-size collet or in a chuck. A workpiece is clamped in the vise, and then it's moved along some axis into the path of the spinning cutter until the desired material is removed. The exact distance that a workpiece is moved in each direction can be dictated by dial scales on each wheel. These scales can be rotated to zero so that you can easily measure the distance moved from any point, like resetting the trip odometer in a car. The speed at which the workpiece is fed might be dictated by hand. On some mills, motorized controls are used, which tends to leave a more consistent surface finish.

That's the gist, but in practice the "speeds and feeds" that were discussed in the "Metal Lathes" entry (page 240) factor heavily into the milling process. And just as with lathes, each bit, cut path, and material will dictate how fast the spindle should rotate and the speed at which the workpiece should be driven into the bit. In addition, there are as many bits for mills as there are for drills, routers, and so on. Each bit has its own limitations. But whatever the bit or speed, cutting oil should be used to help things along. Also, the material should be moved along the side of the bit that feeds into the cutting edge, not unlike the path that should be followed with a wood router. Feeding into the cutting teeth as they spin *away* from the workpiece is called *climb milling* and should be avoided.

If all that seems like a lot to consider, it is. And you can imagine how complicated tool paths and mounting fixtures become when milling something as complex as the head of a car motor. But fortunately, milling can also be approachable. Affordable benchtop mills and a forgiving material like high-density plastic mean that the basics can be learned by trial and error without too much risk to equipment or operator. Give it a go. Even those octogenarian machinists on their World War II–era Bridgeport machines had to get their start somewhere.

Planes

Planes are age-old tools used for flattening wood. The earliest known planes were found in the entombed city of Pompeii and date to 79 CE. What's most remarkable is that those planes look almost exactly like the wooden and steel planes still used in modern Japan. Fast-forward to the iron-bodied planes used in fourth-century CE Rome, and these too are nearly indistinguishable from the metal-bodied planes found in your local hardware store. Planes—at least hand planes—are tools that haven't had to evolve much.

Planes aren't exactly limited to flattening, per se. The intention might also be to reduce the thickness of the wood, remove an old layer, or smooth the surface. Yes, you *can* smooth the top layer with a metal **card scraper**, or with sandpaper, but sharp planes meant for smoothing are capable of cutting a silky texture that can't be beat. To give an idea of the surgical precision that's possible and what sort of glassy finish it must leave behind, look to the hand plane competitions in Japan; the world record for slicing off an unbroken ribbon of cleanly planed wood fiber is 2 microns thick. That's less than half the thickness of a red blood cell!

Precision isn't always the goal, however. Modern planes suit a host of needs and are capable of leaving a textured finish, efficiently truing a piece of lumber, or mulching an entire face of a board in a single pass. What follows are a variety of tools to accommodate these tasks.

Hand Planes

The terminology around **hand planes** can be confusing. They're one of those tools that have existed for so long, in every corner of the globe and across so many trades, that things get a little convoluted. But in general, what is being discussed here are handheld planes used to remove a layer of wood.

Block planes are one-handed devices appropriately sized for a tool belt. They're good for the quick task of shaving some wood off a door so that it will close correctly or knocking down the sharp edges on solid wood and plywood. The bevel on block planes faces up, which runs counter to most other planes. This makes for a stiffer blade, which in turn means that this plane is well suited for the tougher end grain that a carpenter might wish to pare down in the course of squaring up cabinets and windows.

Moving up in size from block planes, we have several larger planes that all tend to be categorized as **bench planes**. These are numbered according to size, and each has its own name. For instance, a **#4 bench plane** is a common multipurpose size that is often referred to as a **smoothing plane**. There are **scrub planes**, which cut hollow gouges, removing a lot of wood quickly. They're often used to leave a rustic scalloped finish, but traditionally you'd look to a **#5 jack plane** after the scrub plane to smooth the surface you just excavated. Jack planes pull double duty, both for smoothing and for leveling, though a true leveling plane (also called a *jointer plane*) is even larger. A **#8 jointer plane** can be as long as an arm and can outweigh a bowling ball. Unsurprisingly, these monsters don't see as much use as they used to, having been displaced by power planers and jointers.

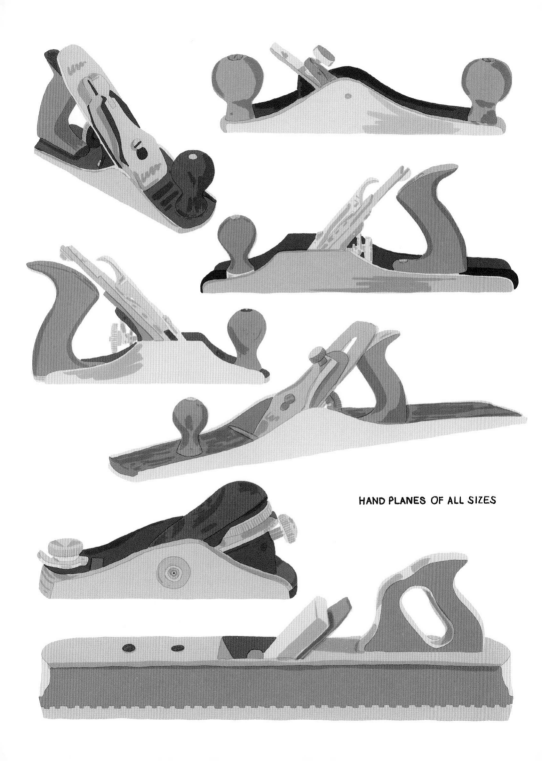

HAND PLANES OF ALL SIZES

Using any of these planes means going *with* the grain to prevent tearout. Holding the plane at a slight diagonal to the direction being pushed will help as well. Apply even pressure, and if using a bench plane with two hands, push down on the front knob and forward with the rear handle. But none of that matters unless these tools are sharp and the plane is set up correctly. The nearly imperceptible difference of a blade protruding too far (or too little) will render any hand plane useless.

Sharpening a plane's blade calls for the same techniques as sharpening a chisel—consult the sharpening section (page 96) for a more detailed process.

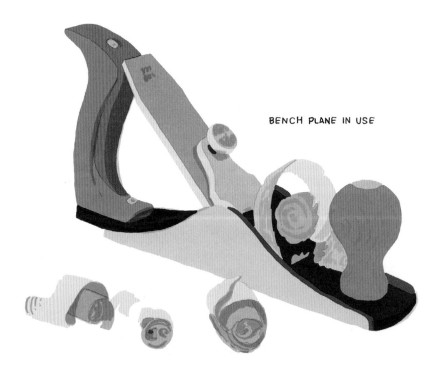

BENCH PLANE IN USE

Once sharp, installing the blade and setting the plane up can be a challenge. The blade (often called the *iron*, harkening back to when it was the only piece of iron in a wooden frame) has a metal plate mounted on top of it, called a *chip breaker*. Set the chip breaker back from the edge of the blade about ¹⁄₁₆ inch [2 mm]. Except with a few models and with low-angle block planes, the iron should be installed bevel down. The blade and chip breaker then attach to a piece of the plane called the *frog*, which is mounted to the body of the plane.

The mouth where the blade protrudes can be adjusted via screws on the body and frog. You can experiment with the size, but in general, a tighter opening is better for finer work. The iron must be mounted so that it's perfectly parallel. If one side of the blade is digging deeper than the other, this defeats the entire purpose of a flattening tool, so there's an adjustment for this as well. Flip the plane over and sight down the bottom of the shoe with the blade protruding. Tweak any lateral adjustments until the blade is visually parallel with the face.

Setting the depth requres practice, and the proper depth depends on the wood. But a basic approach is to retract the blade within the body and run the planeover a piece of wood. Then adjust the blade outward with the adjustment screw, one-eighth of a turn at a time, until finally a push across the wood produces a paper-thin curl of wood. Fine-tuning the depth from there is a matter of personal preference. If there's any noticeable difference in thickness between either side of the blade, recheck that lateral adjustment. And once you're done planing, be sure to retract the blade and wipe down the metal body with oil to prevent tarnishing. Camellia seed oil works nicely.

Power Planers

Electric handheld planers (generally truncated to **power planers**) bridge the gap between the slow-moving and arduous work of handheld planes and the bulky efficiency of freestanding thickness planers. They're capable of removing quite a bit of material, and that potential to quickly overdo it has given these tools a reputation of not being suited for delicate work. However, these tools can be quite accurate with some tuning.

For starters, the blades need to be sharp. On power planers, these blades are usually referred to as *knives*. On some older models, the substantial blades can be removed and sharpened with a set of water stones, using the same techniques you'd use to set an edge on any other blade (see page 93). However, most modern planers use thinner, disposable knives. They're sharp on both sides, so when one edge becomes dull, try flipping the knives over to a fresh edge. Once that's been exhausted, buy a new set. Be sure when flipping them to first clean off any wood pitch or residue.

Next, adjust the knife height. Don't assume that a brand-new planer's knives are correctly adjusted. These should be set both when you purchase the planer and any time you reinstall the knives. The flat surfaces on the bottom of the planer are called the *front* and *rear shoes*. The front shoe is the one that adjusts in elevation with a turn of the knob to vary how much material is being removed. The rear shoe remains fixed, and it's this shoe that the knife height is adjusted against. To make this adjustment, first dial the front shoe as high as it will go, so that it's out of the way. Then flip the planer over so the shoes are facing up.

Rotate the drum (which holds the knives) so that the knife blade is recessed below the top of the drum. Place a ruler on its edge, flat on the rear shoe so that it's suspended over the drum with a gap between the two. Then rotate the drum so that the knife rotates forward and up into the ruler. The knife should gingerly snag the ruler, lifting it and carrying it forward as the blade continues to rotate over the top of the drum. Knife and ruler advance forward and then as the knife slips down over the other side of the drum, the ruler is lowered and is left once again sitting flat on the rear shoe. If the knife height is correctly adjusted, this interaction will have advanced the ruler forward about ⅛ inch [3 mm]. If it moved the ruler less (or didn't snag it at all) the knife needs to be adjusted up. If it advanced the ruler too far, it needs to be adjusted lower. Check both sides of each blade.

Before powering up the planer, first set the front shoe's dial to the lowest setting to remove as little wood as possible. You can adjust up from there, but more passes with a less aggressive bite will leave a smoother face. While planing, take care to keep the planer flat, especially when starting and ending each pass, as the drum can dig into the wood on the ends where it's supported by only half the shoe. Alternatively, you can butt sacrificial pieces of wood against the ends of the workpiece so that the planer is starting and ending each pass on a flat plane.

On just about every power planer, the knives are ever so slightly wider than the base of the planer. Obviously, you should be keeping the planer flat, but this protruding edge means that if you were to tip the planer to its side, the corner of the knife would carve a groove. Some folks like to round off these corners with a stone or a low-speed **bench grinder**. On a wide board that requires multiple passes, you would otherwise be left with some lines that show where the edge of the planer passed by, like lawn mower paths in a yard. Rounding over the corners removes these marks and lessens the need to break out the sandpaper.

Thickness Planers

A **thickness planer** (otherwise known as a **surface planer**, or sometimes just a **planer**) is a freestanding power tool that has a wide set of blades to remove material from the entire width of a board as the wood is fed through the machine.

Thickness planers differ from other planes and planers in that they are referencing the bottom of the board and shaving away wood from the top. The goal when feeding a board through one of these planers—perhaps several times—is that it will eventually emerge out the other side both with a fresh face and a consistent thickness. **Drum sanders** are less common and remove less wood, but they work much the same way, creating a fresh surface atop a board of consistent thickness as it's fed through.

The bit about the consistent thickness is important. A board that is curved will pass through a thickness planer and still be curved; this is not a planer that flattens a board. To produce a flat board of consistent thickness, you first need to create a flat face on a jointer. Then you can run the board through the planer, flat side down, creating a parallel and flat face on the opposite side.

Before use, a planer must first be adjusted to the general height of the board that's about to be fed through, plus a little tighter so that the bladed drum can shave off the desired amount. Most have a depth gauge that works by lowering the planer over the wood until the gauge indicates the amount to be removed. Don't try to remove more than ⅛ inch [3 mm] in one pass. And even that would be the absolute limit of most planers and wood types; the wider the board, the less you should be trying to remove. So, as with other planers, less is more. Start small and send the board through a few times if need be.

Many planers have a switch to select between dimensioning or finishing the wood, which is an adjustment of blade speed for a rougher or smoother surface. Some also have a depth stop; reducing the thickness of a board might involve a half dozen passes and height adjustments, and this depth stop will prevent you from accidentally overshooting the final mark.

The way a planer operates is that feeding rollers grab the wood as it's fed in, sending it under the planer knives and out the other side. (These rollers also compress the wood, which is why an unjointed and cupped board can be sent through and planed "flat," yet will spring back to being cupped when it exits.) The beginning and end of this process can result in planer *snipe*, which is when the planer knives dig deep on each end of the board because it's only supported by one feeder roller. This can be avoided by supporting the far side of the board as it both enters and exits the planer. Or just don't feed through boards that are cut to their final length. Plane their thickness first, planning for a little snipe, and then trim the ends to size.

Rasps

Rasps are similar to files (see page 231) in that they are abrasive, hardened-steel hand tools meant to remove material. They're limited to work on softer substances, such as wood, plastics, and leather, and they're even used on horse hooves by hoof-care specialists known as *farriers*. (Appropriately, their rasp of choice is a **farrier's rasp**.) Rasps remove material much more quickly than files because they have separate triangular teeth (as opposed to a file's parallel lines), and those teeth individually cut deep grooves, scraping away at a volume that files can't compete with. This option is essential in woodworking, where shaping can require an aggressive sculpting tool.

Like files, rasps come in a few degrees of coarseness and have equally cryptic names to describe them. General-purpose rasps are the coarsest. You may see them called **bastard rasps**, not unlike the files of the same name. **Cabinet rasps** (also known as *second-cut rasps*) have medium coarseness and tend to be the most common. **Patternmaker's rasps** are smoother and leave the finest surface of all the rasps, but the work will be slower. And just like files, rasps are also available in a variety of cross sections: flat, half round, round, and triangular. **Cranked-neck rasps** have a kink in the middle that allows for work in confined spaces, and **curved rasps** are designed for, well, curves.

For smaller projects and tight corners, such as wood carving or instrument making, **rifflers** are used. *Riffler* comes from the old French word *rifler*, meaning "to file" or "scrape." The coarseness of these tools is graded using the same Swiss pattern used with some files. A riffler is basically a small, double-ended rasp with a gripping area in the middle. It can be straight or curved and is available in the same assortment of profiles as the larger rasps.

A different style of rasp that stands on its own and is possibly of more use than all the others is the **saw rasp**. Unlike all the previous models, which are lengths of solid steel, saw rasps (most notably manufactured by Shinto) have a number of saw blades arranged in a diamond pattern that looks like the high-traction grating of factory walkways. These rasps will plow through material and allow you to swiftly shape a piece of wood to fit any need. They are aggressive in their removal of stock but also controllable, making them a useful piece of kit for any builder.

When using any rasp—be it traditional, riffler, or saw rasp—hold it diagonal to the direction being pushed, and use the full length of the rasp. Lift the rasp off the workpiece after each push so that you're cutting on the push stroke only. Putting pressure on the pull stroke will dull the teeth. When using a round or half-round rasp, rotate it as you push to use the full breadth of the teeth. Use a longer rasp than you think you need. As with files and sandpaper, work from coarsest to finest. Take care not to hog out too much material, which can easily happen. And clean the teeth of your rasp with a **natural-bristle brush** once the job is done.

Rotary Tools

Rotary tools (also known as **die grinders**) are handheld power tools that spin a wide variety of bits and attachments at high speeds upward of 20,000 rpm. Given the popularity of the brand that Albert J. Dremel brought to market in 1935, they're often broadly referred to as **Dremel tools**, regardless of the actual manufacturer.

Rotary tools serve two needs. First, they can get into hard-to-reach spots and accomplish nimble tasks where larger tools can't compete. Tiny cutoff wheels or grinding attachments can be just what's required to remove an elusive weld or clean up a fabricated bit of metal. Second, their small footprint and vast array of attachments have made them hobbyist staples. A maker in an apartment can do small-scale woodworking, metalworking, costume and prop fabrication, jewelry making, and any number of other crafts with hardly more than a rotary tool and a large collection of tiny attachments. In fact, fabrication at this scale is sort of its own niche, and the German manufacturer Proxxon (which makes a fine rotary tool of its own) has an entire line of **micro-benchtop tools**: a table saw, lathe, drill press, and more, each small enough to fit on a dinner plate.

To understand what a rotary tool is capable of requires a look at its attachments. There are sanding drums of different sizes with replaceable sandpaper, and disposable flap discs; stone-grinding bits of various shapes; and abrasive diamond bits that look like those you'd see a dentist use. (In fact, a dentist's drill is a specialized rotary tool with a flexible shaft.) Coin-size cutoff wheels like those found on an angle grinder are frequently used. There are also cutting bits that wrap helical teeth around different profiles, and these can be used like a router bit. And if tiny circular saw blades raise the intimidation factor of what otherwise feels like a somewhat harmless tool, cotton buffing wheels are there to lower it. And those are just the accessories that might come in a basic kit. More complex accessories include planer attachments and sharpening attachments. There's even a workstation that can be used with many brands of rotary tools where the tool can be mounted at any angle to shape workpieces with the control of two hands. This mount can also be plunged like a drill press. Plus, if you set the depth to table height and attach a sanding drum, you now have a pint-size drum sander.

Given all that adaptability, it's no wonder these tools often satisfy the needs of someone whose fabrication aspirations exceed their storage space. Rotary tools are not necessarily the best tool for the job, but often the one that will suffice. Conversely, that same adaptability and one-handed pen-like precision mean that, on occasion, a rotary tool is exactly what you need.

Routers

Routers are woodworking power tools that hollow out recesses and shape edges and corners by sitting flat on a workpiece with a rotating cutter protruding straight down into the workpiece. They're available as hand tools known as **router planes**, but these are used infrequently and only by the fine woodworkers who appreciate their delicate touch. Router planes are also far less capable than the more common power tool variety.

You'd be excused for overlooking how much influence routers have had on the wood within your home: the rounded edges of a table, the artistic profile of a picture frame, the contours of trim, and the texture of a cabinet door panel. The same goes for what's outside your walls: exterior trim and door casing, the beveled edges of a picnic table. Even those wooden signs at the park, with their recessed lettering filled with yellow paint, were carved with a router.

Electric routers all work the same: by spinning a router bit at high speed, up to 35,000 rpm. To wrap your head around how fast 35,000 rpm is, consider that a blink of an eye happens in about a tenth of a second. So in the time it takes you to blink your eyes just once, that bit has spun around almost sixty times. It's fast. Not that all bits can or should be run this fast. The velocity of a rotating spindle is an important consideration with routers and a critical calculation in milling and lathe work, which takes into account bit diameter and the speeds appropriate for the material being cut.

Palm routers—often cordless—fit into a single hand and spin smaller bits. They're sometimes called **trim routers**, as they're often used to cut back overhanging veneers or for the finish work of rounding over an edge. **Midsize and full-size routers** are two-handed machines capable of performing more powerful work and spinning larger bits, like those that might be required to create molding profiles or to flatten an entire slab of wood. These routers can all be had with a plunge-style base that allows the user to press a bit down into wood—a critical feature for hollowing out mortises or carving text into a sign.

Rotational speeds are all about achieving a certain linear velocity for the cutting edges that spin around. Smaller-diameter bits cover less distance because of their smaller circumference, so their blades must move faster to match the linear velocity of the blades on larger diameter bits. What's linear velocity? Think of it as if you were a tiny human standing next to the spinning bit: It's the speed at which that cutting edge flies past you, like a passing car.

For example, the cutting edges of a 1-inch [25 mm] diameter router bit spinning at 35,000 rpm will have the same linear velocity as a 2-inch [50 mm] bit spinning at only 18,000 rpm—approximately 100 miles per hour [160 kph]. If you scale that bit up to the diameter of a truck tire, you'll achieve the same linear velocity with a comparatively meager 1,200 rpm. The danger is if that truck tire was spun at our original 35,000 rpm, its linear velocity would destroy the tire (and the truck). So, diameter matters with rotational speed. Take note when your router specifies the largest bit it can spin.

Then there are **router tables**. A full-size router is mounted to the underside of the table, with its adjustable bit protruding up into the air. Rather than moving the router along a workpiece, the user manipulates the workpiece against the stationary bit. Router tables can be purchased, though lots of folks build them. With a router table outfitted with a guide fence, carving perfect slots or routing long channels is a breeze.

Lastly—and these are far less common but worth mentioning—we have computer-controlled routers, known as **CNC routers**. Where these really shine is in cutting intricate shapes into sheet goods. A lot of folks have designed templates for plywood furniture that requires no mechanical fasteners or glue, and these machines can churn out the pieces from a full sheet of plywood in no time at all. Of course, CNC routers that are capable of working on a full sheet of plywood are roughly *the size of a full sheet of plywood*, but their cost, applications, and approachability still make them more attainable and useful to hobbyists than other CNC tools like mills or lathes.

Types of routers aside, what matters far more is the bit. Not unlike a rotary tool, the magic of a router isn't that it spins, it's *what* it spins. There are hundreds of types of router bits with different profiles, and many of these have ball bearings that guide the cutting edge. Here are a few bits and how they work.

Roundover bits are run along the edge of a piece of wood to turn a square corner into a round one. The lumber you buy from the big box store typically has rounded edges, but the end cuts are not rounded, so if you were to use this wood for decking, or for the handrail of that deck, grabbing a palm router and rounding over the

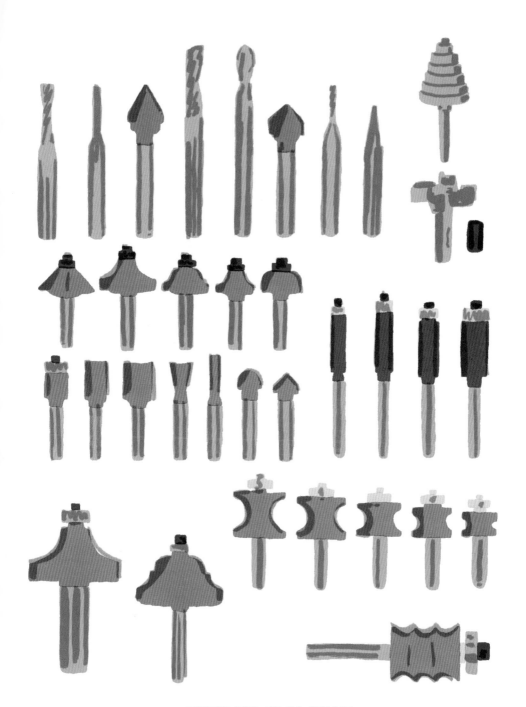

ROUTER BITS OF ALL SHAPES

edges of the end cuts would go a long way toward creating a polished look. **Rabbeting bits** are used to cut a shoulder in the edge of a workpiece, often to join two pieces. These have interchangeable bearing diameters, so that the depth of the cut can be adjusted. **Flush trim bits** are used to trim an overhanging veneer flush with the rest of the wood. **Roman ogee bits** are decorative bits that cut wavy profiles along the edges of molding and picture frames. There are **straight bits**, too, which look like drill bits with elongated flutes. These are most often used to plunge into a surface and cut a slot. Because they are plunging and moving freely about, often no bearing is attached, so a jig or router table would be needed to keep the slot straight. These aren't the only bits without bearings: The broad heads of **planer bits** are used on sled-held routers to flatten the tops of large wood slabs that wouldn't otherwise fit into a planer.

When it comes to using these bits, there are a couple of important rules to follow. Route with the grain if possible. If the bit is digging under and against the grain, you'll get tearout. In that same vein, when routing all the way around a piece, do the end grain first, as the corners will tear out when you reach the long grain. Routing the long grain after the tearout on those corners will remove this mess. Also, don't power through a cut at full depth. Doing so with a slight roundover is one thing, but where more material is being removed, do it in a few passes at staggered depths until the full cut is achieved.

Most important is to understand the direction of the feed. You want the wood to be fed into the cutting edge. When looking down at a router from above, the bit is spinning clockwise, which is usually indicated with an arrow on the router. That direction of bit rotation means that when moving around the outside of a piece of wood (which is probably most common) you want to move the router in a *counterclockwise* direction. If you're routing *within* a piece of wood, this movement is reversed, and you want to move the router in a *clockwise* direction.

Let's say you're making a giant plywood donut as wall art for your new donut shop. To round over the edges on the *inside* of the donut's hole, with the router *within* the workpiece, the router is moved clockwise around the edge. To round over the outside edge of the donut, with the router on the *outside* of the workpiece, you move it in a counterclockwise direction.

Also, congratulations on your donut shop.

ROUTER FEED DIRECTION

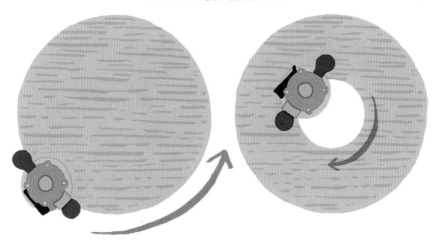

Sanders

Sanding is the process of using an abrasive to remove a layer of material while scuffing or polishing the surface. To hobbyists and those in the world of fabrication and construction, the term *sanding* is reserved for handheld and machine-powered sandpapers, but the principles hold true from the largest to the smallest abrasives—whether it's water and sand polishing river rocks, a whetstone sharpening a chisel, or a buffing wheel bringing out the shine on a car fender. It's all the same.

Sanding is generally done in stages because larger abrasives will only get you so far, and relying only on finer abrasives to produce a glossy finish could take eons.

To understand sanding and sanding tools, you first need to get to know sandpaper intimately.

Sandpaper

Sandpaper owes its name to the sand and glass abrasives that were once used to make it, but modern versions use other materials, such as garnet, aluminum oxide, and silicon carbide. To create sandpaper, drums of paper (or cloth) are coated in resin and drawn across trays of abrasive particles that have been statically charged. The particles leap onto the paper and get embedded in the resin, which then cures to cement a durable and rough surface on one side of the paper.

The roughness of the sandpaper, or its *grit*, is determined by the size of the abrasive bonded to it. You'll find the grit printed prominently on both the packaging and the back of the paper. In most systems, the smaller the number (say, 160), the coarser the grit. The higher the number (say, 220), the finer the grit. In the United States and several other countries, the CAMI system (named for the Coated Abrasives Manufacturers' Institute) is used, and its numbers are derived from the number of holes per inch in the screens that sift the grains during manufacturing. But because manufacturers use a couple of different grading systems (and these might be for sale in the same store), you'd do well to know their differences.

European manufacturers have their own system of grading: FEPA (named for the Federation of European Producers of Abrasives). FEPA sandpapers are easily identified because their grades start with the letter *P*—for example, P180. Fortunately, the FEPA system and the CAMI system are fairly comparable through the most commonly used sandpaper grits, but above P240, things start to diverge. The 320-grit CAMI is approximately equal to P400, 440-grit CAMI is roughly P600, and 800-grit CAMI is equal to P1500. There's also the Japanese micron system, which runs counter to the other two grading systems, where larger numbers indicate a coarser grit.

An immense variety of grits are available, but the truth is that most people only need a handful. A piece of 80-grit sandpaper is great for shaping wood and smoothing a joint. It will round over edges and strip paint. A sandpaper with 120 grit will do all that as well and will produce a smoother finish, but it'll require more work. Sanding wood smooth enough for an oil finish might call for

220 grit, but anything beyond that won't do much to improve the final product. The same 220-grit sandpaper will also work between coats of polyurethane or for removing rust from metals, and when applying something like lacquer, a light sanding with 220 before the final coat will guarantee a smooth finish.

Understanding the grit numbers of sandpaper is probably the most important aspect of sanding, but knowing which types of sandpaper are best for particular applications isn't to be overlooked.

Aluminum oxide sandpaper works well for general sanding of woods and metals and will stay sharp longer because the abrasives fracture to expose new edges. **Ceramic red sandpaper** is more durable and costly but of great value when attached to power tools. **Garnet sandpaper** has a softer brown abrasive that wears quickly and is usually made with cheaper paper. **Silicon carbide sandpaper** is often blue or black and uses waterproof paper. It can be used on wet or dry surfaces, and the water will help wash away particulates and keep the paper from loading up with dust. It's an ideal sandpaper for metals or plastics or for polishing between coats of finish.

When it comes to technique, sanding with the right type and grit of sandpaper can be as simple as holding a piece in your hand and working it in random circular patterns. But the pressure applied along your fingertips will be uneven, so if you're trying to create a flat surface, you'll want the sandpaper to be flat. A common solution is to wrap the paper around a block of scrap wood (or another object) and then work it like a scrub brush. **Sanding blocks** are tools that hold sandpaper, fit in the palm, and offer nimble corners to access nooks and crannies, but to take this evolution a step further, manufacturers now offer sanding blocks that are themselves abrasive. These malleable blocks and wedges look like a kitchen sponge and come in the same grits as their sandpaper brethren.

Random Orbital Sanders

You'll sometimes hear **random orbital sanders** called **palm sanders**, given their size and easily gripped dome top. *Palm sander* is actually a name that's thrown around for a few styles of powered sanders that are all held with one hand. The original palm sander (of the nonorbital variety) is a small device with a rectangular pad that uses a quarter sheet of sandpaper. Perhaps more confusingly, the larger models that use a half sheet are officially labeled as **orbital sanders**. Then there are **mouse sanders**, which have a pad shaped like a clothing iron and are useful for tight corners, and **straight-line sanders**, which have a V groove for working edges. All of these fit in your palm and so are often called *palm sanders*. It's confusing, to be sure. But these days, the tool that people will most often grab when on a mission for a handheld palm-gripped sander is the random orbital sander.

The modern random orbital sander eschews clipped-on rectangular sheets of sandpaper in favor of custom sheets that adhere to a round pad. The biggest thing to get right when matching sheet to sander is the hole pattern on the bottom. Both the sander pad and the sheets have holes meant for dust extraction—either into an attached bag or a **dust-collection vacuum**—and if the holes don't line up, dust will build up under the pad and gum up the works.

When using the sander, don't apply too much pressure, as the pad won't be able to oscillate. Also, if you spend too much time in one spot, the sander might leave a noticeable mark, especially around the edges of the pad. Keep the sander moving, and allow the weight of it to do most of the work. Change your sandpaper as often as is needed, rather than wearing through and damaging the underlying (but replaceable) pad. And if you notice the sandpaper disc spinning freely on a well-used sander, the brake pad is likely worn out and should be replaced.

Belt Sanders

Handheld **belt sanders** can be pugnacious beasts. They buck when the power is flipped on and that broad treadmill of sandpaper ramps up to speed. It's jarring. They have an earned reputation for aggressive work, often replacing a planer in removing layers of wood from flat surfaces. They are an ideal tool for resurfacing beams or floors, though freestanding **floor sanders** are also an option for the latter. But that same power can be both an asset and a liability. Belt sanders work fast. By rotating through a full belt of sandpaper, they are working with far more surface area, so the paper stays abrasive for longer. This means that staying in one place for too long, or digging in with the edge of the sander, will yield immediate results and likely immediate regret.

When working with a belt sander, select lower speeds when possible. Keep the sander flat, especially over edges, which can be quickly rounded. Butting up a sacrificial piece of material that's the same height can help keep this from happening. Gouges will start to form if the paper gets loaded with grit or wood pitch, so keep an eye on that and use a clean belt. But mostly, when it comes to using a belt sander, the key is to reserve it for the work that calls for it. There's no sense using a grit finer than 120, or for pulling it out to do any finish work. Stick to truing up faces, stripping old varnish, or making a first pass across a rough-sawn board. If you keep to these types of jobs, you'll see the value in this tool, rather than its drawbacks.

One alternative use—and a way to gain more control over a belt sander for some finer work—is to turn it upside down and clamp it to a workbench, leaving the spinning belt exposed; you'll have two hands free for manipulating a workpiece across it. With a secure clamp and the usual eye and ear protection that should accompany sanding, this is just as safe as using the belt sander the way it was intended. However, if you intend to make a habit of using your belt sander this way, a benchtop sander might be what you need instead.

Benchtop Sanders

Benchtop sanders (and their larger, freestanding cousins) operate as workstations rather than handheld tools and come in a few packages. There are belt sanders that spin a flat belt horizontally, or vertically, and these might have a table adjacent to the belt. There are also **disc sanders**, which rotate a plate of sandpaper perpendicular to a worktable. Most of these tables are adjustable so that different angles can be sanded, and many have a built-in miter gauge. In fact, all of these features are often combined in a tidy package that's sold as a combination sander.

The flat surfaces of benchtop sanders are good for, well, flat surfaces. But the table on a disc sander is also ideal for sanding

FLAT FILE SANDING STICKS

DETAIL SANDER

DISC SANDER

OSCILLATING SPINDLE SANDER

SANDPAPER HOLDER

BELT SANDER

RANDOM ORBITAL SANDER

outside curves. You can pivot a roughly sawn workpiece on the table and across the disc, knocking down the sharp corners until all the edges are rounded. But what about inside curves? Some of the aforementioned belt sanders have free-hanging belts that can be worked around inside curves, but in general, sanders with a flat belt or flat disc are ill-equipped for the job. For this, you need the roundness of an **oscillating spindle sander**.

Oscillating spindle sanders are benchtop (or freestanding) electric sanders that spin a pedestal of sandpaper while pumping it up and down to vary the path of the abrasives. These are the quickest way to sand inside curves, but there are alternatives.

Some folks will wrap sandpaper around a dowel to sand interior curves by hand. You can also attach the dowel to a drill for faster results, or glue sandpaper to any other drill-attached drum, such as a retired hole saw. Any of these methods will work just fine, and the same rules apply regardless of what you end up using to do the sanding: Move through grits in a progression from coarsest to finest. Use varied fluid movements across a clean surface. And keep the aggression to the minimum amount required as you refine the workpiece. All of this should seem obvious; the through line of this refinement process is intrinsic to blade sharpening, filing wood, grinding metal, and really all of shaping.

As with all tools, the principles are universal, and there's no reason to get mired in the minutiae once the fundamentals become intuitive. As the famously undefeated Japanese swordsman and philosopher Miyamoto Musashi wrote in 1643, "If you know the way broadly, you will see it in all things."

INDEX

Acknowledgments

Just as I finished writing this book, our property—along with every tool I owned—burned in the fires that raged across California in late 2020. It was a tragic loss, but the most important elements survived: the experiences shared and the knowledge gained.

I had the privilege of learning quite a bit alongside some amazing people, the most important of whom is my partner, Molly. This book wouldn't exist without that property, and that property would not have existed without her. For that, for her, and for every other good thing in my life that she's made possible, I'm eternally grateful. Thanks as well to the many dozens of friends who showed up time and again to lend tools, lend a hand, and make that place and community what it was. (And special thanks to all those who looked after me when I broke my back in an ill-fated attempt to use a borrowed brad nailer on our unfinished cabin loft.)

I also owe a debt of gratitude to the specific individuals who, over the years, furthered my interest and education in tools or in writing or who otherwise contributed to this book in some fashion. Those folks (and I'm certainly forgetting many): Tom Bonamici, Scott Braun, Krishna Bulkin, Jeff Canham, Ben Coddington, Ben Corman, Jimmy Diresta, Tyler Fayles, Michael Getz, Drew Hartley, Jason Hemmerle, Greg Hennes, Rachel Hiles, Joey Hiller, Mel Ho, Charlie Hoehn, Ryan Holiday, Paul Jackman, Lloyd Kahn, Deanne Katz, Zach Klein, Jess Kokkeler, Tucker Max, Tim McSweeney, Charlie Miller, Thomas Murray, Darren Nakata, Nils Parker, Mikey Rosenzweig, Carey Smith, Michael Soloway, Ashley Tackett, Sebastien Tilmans, and of course, Dick Proenneke.

JEFF WALDMAN is a maker, builder, and creator with a talent for finding interesting people and even more interesting projects. He is currently designing and building a cabin and outdoor retreat space in Santa Cruz, California, and tinkering in his shop in Oakland, California, but he's always looking for his next project.